Sound, Space and Civility in the British World, 1700–1850

This edited collection examines the sonic dimensions of civility in the British world in the eighteenth and the first half of the nineteenth centuries – what might be called the long Enlightenment. The enormous emphasis on civility during this period produced new ways of listening and new meanings of sound and noise. Through an evaluation of the impact of shifting understandings of sound and noise, the essays in this collection restore the acoustic aspects of civility, exploring how censured behaviours and recommended practices resonated beyond the written word. In addition, they show how conflicts over what counted as desirable or undesirable sound were influenced by the spaces in which such sound was heard and generated, from arenas of urban sociability and religious enthusiasm to theatres of colonial encounter and political protest. While noise was frequently criticised from the perspective of civility, however, it was not always perceived as injurious. As the essays in this volume demonstrate, noise was heard in all sorts of positive ways as a crucial component of life in a modern, commercial nation and its empire. Changing distinctions between sound and noise sometimes challenged existing understandings of civility and at other times licensed new forms of counter-civility. Hearing the sonic dimensions of these diverse types of civility across a range of locations and contexts, this edited collection places sound at the centre of a new approach to the literature and history of Britain during an important era in the transition to modernity.

Peter Denney is a Senior Lecturer in History at Griffith University.

Bruce Buchan is Associate Professor of History at Griffith University.

David Ellison is a Senior Lecturer in Literary Studies at Griffith University.

Karen Crawley is a Lecturer in the Griffith Law School at Griffith University.

British Literature in Context in the Long Eighteenth Century
Series Editors: Eugenia Zuroski Jenkins and Jack Lynch

This series aims to promote original scholarship on the intersection of British literature and history in the long eighteenth century, from the Restoration through the first generation of the Romantic era. Both "literature" and "history" are broadly conceived. Literature might include not only canonical novels, poems, and plays but also essays, life-writing, and belles lettres of all sorts, by both major and minor authors. History might include not only traditional political and social history but also the history of the book, the history of science, the history of religion, the history of scholarship, and the history of sexuality, as well as broader questions of historiography and periodization. The series editors invite proposals for both monographs and collections taking a wide range of approaches. Contributions may be interdisciplinary but should be grounded in sound historical research. All manuscripts should be written so as to be accessible to a wide audience while also making lasting contributions to the field.

Invoking Slavery in the Eighteenth-Century British Imagination
Srividhya Swaminathan and Adam R. Beach

Making Gender, Culture, and the Self in the Fiction of Samuel Richardson
The Novel Individual
Bonnie Latimer

Restoration Stage Comedies and Hollywood Remarriage Films
In Conversation with Stanley Cavell
Elizabeth Kraft

Sound, Space and Civility in the British World, 1700–1850
Edited by Peter Denney, Bruce Buchan, David Ellison and Karen Crawley

For a full list of titles in this series, please visit www.routledge.com/series/ASHSER-2115

Sound, Space and Civility in the British World, 1700–1850

Edited by
Peter Denney, Bruce Buchan,
David Ellison and Karen Crawley

Routledge
Taylor & Francis Group
LONDON AND NEW YORK

First published 2019
by Routledge
2 Park Square, Milton Park, Abingdon, Oxon OX14 4RN

and by Routledge
52 Vanderbilt Avenue, New York, NY 10017

Routledge is an imprint of the Taylor & Francis Group, an informa business

British Library Cataloguing-in-Publication Data
A catalogue record for this book is available from the British Library

Library of Congress Cataloging-in-Publication Data
A catalog record has been requested for this book

ISBN: 978-1-4724-6659-4 (hbk)
ISBN: 978-1-315-60994-2 (ebk)

Typeset in Sabon
by codeMantra

MIX
Paper from responsible sources
FSC® C013056
FSC www.fsc.org

Printed and bound in Great Britain by
TJ International Ltd, Padstow, Cornwall

Contents

List of figures

List of tables

List of contributors

John Barrell is Emeritus Professor of English at Queen Mary University of London and an honorary fellow of King's College, Cambridge. He has written numerous books on British art, literature and politics in the eighteenth century and the Romantic period. His most recent work is *Edward Pugh of Ruthin, 1763–1813: 'A Native Artist'* (University of Wales Press, 2013).

Bruce Buchan is Associate Professor of History at Griffith University. He is the author (with Lisa Hill) of *An Intellectual History of Political Corruption* (2014) and *The Empire of Political Thought: Indigenous Australians and the Language of Colonial Government* (2008). He has been appointed to visiting professorships at the University of Copenhagen (2015–16) and the École des hautes études en sciences sociales, Paris (2017). He is currently working on a new project on the conceptual prehistory of race entitled 'The Borders of Humanity: Linnaean Natural Historians and the Colonial Legacies of Enlightenment' (with Linda Andersson Burnett) funded by the Riksbankens Jubileumsfond in Sweden.

Karen Crawley is a Lecturer in the Griffith Law School. She researches in legal theory and popular culture, with an interest in the theatricality of law and legal processes, and feminist and aesthetic approaches to justice. She holds a Bachelor of Arts/Law from the University of Sydney and a doctorate in Comparative Law from McGill University. She is Managing Editor of the *Australian Feminist Law Journal* and active on the board of the *Law, Literature and Humanities Association of Australasia* (lawlithum.org).

Peter Denney is a Senior Lecturer in History at Griffith University, where he was recently a recipient of an Australian University Award for Teaching Excellence. He has published essays on a range of topics, including plebeian poetry, popular radicalism, landscape gardening, the rural soundscape, ideas of poverty and the history of the senses. He is currently completing a monograph on landscape and soundscape in Britain in the long eighteenth century.

Markman Ellis is Professor of Eighteenth-Century Studies at Queen Mary University of London. He is the author of *The Politics of Sensibility* (1996), *The History of Gothic Fiction* (2000), *The Coffee-House: A Cultural History* (2004), co-author of *Empire of Tea* (2015) and editor of *Eighteenth-Century Coffee House Culture* (2006) and *Tea and the Tea-Table in Eighteenth-Century England* (2010). He is currently working on sociability and intellectual culture in London in the mid-eighteenth century.

David Ellison is Senior Lecturer in the School of Humanities, Languages and Social Science at Griffith University. His research focusses on the literary and cultural histories of Victorian domesticity. He recently published *On Discomfort: Moments in a Modern History of Architectural Culture* (2017), with Andrew Leach.

Helen Groth is Professor of English and Associate Dean of Arts and Social Sciences, University of New South Wales. Her publications include *Victorian Photography and Literary Nostalgia* (2003), *Moving Image: Nineteenth-Century Reading and Screen Practices* (2013) and, with Natalya Lusty, *Dreams and Modernity: A Cultural History* (2013). She has recently co-edited two collections, *Sounding Modernism* (2017) and *Mindful Aesthetics: Literature and the Science of Mind* (2014), and has published on Victorian visual culture, literature and dreams and, more recently, sound and literature. She has two current book projects, one on 'The Listening Reader' and another on rioting in the literary archive from 1780 to the present.

Associate Professor Dolly MacKinnon is a social and cultural historian at the University of Queensland. Dolly is known for her work on mental, physical and auditory landscapes, with a focus on the historical soundscapes of early modern bells, charity, children and madness.

Gillian Russell is Professor of Eighteenth-Century Studies at the University of York, having taught previously at the Australian National University, Canberra, and the University of Melbourne. Her books include *The Theatres of War: Performance, Politics, and Society 1793–1815* (1995) and *Women, Sociability, and Theatre in Georgian London* (2007). A study of the history of printed ephemera entitled *The Ephemeral Eighteenth Century: Print, Sociability and the Cultures of Collecting* is forthcoming with Cambridge University Press.

Penny Russell is Bicentennial Professor of Australian History at the University of Sydney. Her research spans gender, exploration and the social relations of settler colonial society in the British world. Recent books include *Savage or Civilised? Manners in Colonial Australia* (2010) and (with Nigel Worden) *Honourable Intentions? Violence and Virtue in Australian and Cape Colonies, c. 1750 to 1850* (2016).

Mark Smith is Carolina Distinguished Professor at the University of South Carolina, Columbia. He is the General Editor of Interdisciplinary Studies in Sensory History published by the Pennsylvania State University Press. His most recent work is *Smell and History: A Reader* (West Virginia University Press).

Richard Yeo is an historian of science, a Fellow of the Australian Academy of the Humanities and Emeritus Professor at Griffith University. His books include *Defining Science: William Whewell, Natural Knowledge and Public Debate in early Victorian Britain* (Cambridge, 1993); *Encyclopaedic Visions: Scientific Dictionaries and Enlightenment Culture* (Cambridge, 2001) and *Notebooks, English Virtuosi, and Early Modern Science* (University of Chicago Press, 2014). He has co-edited special issues of *Parergon* (2009), with Vanessa Smith, and of *Intellectual History Review* (2010), with Ann Blair.

Acknowledgements

Peter Denney, Bruce Buchan and David Ellison would like to express their deep thanks to the Australian Research Council for funding the research which resulted in this book. We would also like to convey our appreciation to the scholars who participated in our initial symposia in 2014 and 2015 and to the referees who gave up their time to review chapters. In addition, all the editors are immensely grateful to Dr Eleanor Morecroft for her superlative assistance in preparing the manuscript for the publisher. Many thanks, too, to the Griffith Centre for Social and Cultural Research for supporting our work on the history of noise over several years. Finally, we greatly appreciate the faith that Routledge has shown in this project, along with the patience and professionalism of the editorial team, who have skilfully overseen its publication.

Acknowledgements

Introduction

Listening to civility

Peter Denney, Bruce Buchan, David Ellison, and Karen Crawley

While civility has long been recognised as a central preoccupation of Enlightenment culture in Europe, it has rarely been considered in the acoustic contexts in which it was performed, represented, discussed and heard. This book focusses on how ideas of civility were shaped by changing experiences and meanings of sound in the British world during the long eighteenth century. In 1709, when the English moralist Lord Shaftesbury considered the best way to profit by the studied appreciation of poetic genius, he claimed it lay in contemplative 'retirement' and 'seclusion':

> What Relish then must the World have (that common World of mix'd and undistinguish'd Company) without a little Solitude; without stepping now and then aside, out of the *Road* and *beaten Track* of Life, that tedious Circle of Noise and Show, which forces weary'd Mankind to seek relief from every poor Diversion?[1]

In Shaftesbury's world, the gentlemanly pursuit of civility required a retreat from the wearisome and noisy distractions of a turbulent world of undistinguished company, even if the purpose of such temporary withdrawal was to prepare a polite individual to responsibly 'relish' refined sociable activity. From this perspective, noise is a phenomenon which disturbs the peace and ruins the stillness sought by the gentleman who epitomises civility through his good breeding, discerning taste and elegant manners. Indeed, as Hillel Schwartz has reiterated, noise 'disrupts the initial stillness' and is 'forever conditioned by nostalgic yearnings for a calm that is no more, pleadings for a peace to be restored, oracles of a harmony yet to come.'[2] If noise was, for the aristocratic Shaftesbury, merely the wearying consequence of one's immersion in an increasingly undifferentiated commercial society governed by distraction and display, noise has now become, even more, the acoustic imprint of frantic modernity. Modernity echoes in the thronging hubbub of crowded cities, the impatient roar of incessant traffic, or the imperious din of busy machinery. As a result, Howes and Classen suggest that Western modernity has been shaped by the ceaseless quest to escape from, to regulate or to mitigate 'auditory nuisance.'[3] Slightly differently, Schwartz argues that our modern sonic sensibility is driven by the art of 'listening *through*' the 'elemental' noises that threaten to screen or silence more purposive and integrative sound.[4]

Whether by listening through discordant sonic backgrounds, or by nostalgically yearning for a shattered stillness, Enlightenment, Romantic and Victorian commentators tended to consider purposive and meaningful sound as the acoustic opposite of random and intrusive noise. Sound – of speech or music – was that which enabled and adorned civil life, while noise threatened its extinguishment. In Shaftesbury's gentlemanly ideal, the sonic adornments of civil conduct were also audible privileges of rank – of education, leisure and especially the carefully modulated cadences of polite and agreeable conversation.[5] As Emily Cockayne has argued, British discourse in the eighteenth century was marked by an increasingly strident and elitist sonic normativity. Noise was the non-conversational sonic surplus emanating from the poorer classes and insistently complained of by those in 'polite' society.[6] This prescriptive sonic normativity echoed, for example, the recommendations of Daniel Defoe in 1697 to 'polish and refine the *English* Tongue' by eliminating the 'hard words and long and unusual Coupling of *Syllables* and Sentences' he believed were corrupting the 'civility' of English discourse, making it 'harsh and untuneful to the Ear.'[7] Such refinement of conversation was, according to Peter Denney, part of a broader reclassification of plebeian speech and social activity as undesirable noise, a development which led, among other things, to the segregation of rural as well as urban sensory spaces.[8]

Harsh and untuneful noise, however, was the inescapable audible accompaniment of the rapid expansion of Britain's commercialised and increasingly consumerist urban spaces in the eighteenth century.[9] Its very ubiquity, Cockayne argues, meant that firm distinctions between discordant or déclassé noise and polite or pleasing sound were not always made in eighteenth-century British discourse.[10] The term 'noise' may simply have been appended to any sound 'out of place,' to borrow the influential formulation of Peter Bailey.[11] What determined the proper place of sound, or of noise, draws our attention to the integrative force of the spatial characteristics of sonic phenomena. Alain Corbin, for example, has explored how the intonation of village bells in nineteenth-century France constituted an 'auditory space' sonically 'conjure[d] up' by the metre of their reverberant sound.[12] Conflicts over sound, Corbin demonstrates, necessarily entailed disputes about the meanings and uses of space, along with its role in constructing social, regional and national identities. More recent scholarship has emphasised the role of multisensory experience in shaping the perception of and participation in public spaces in the eighteenth century.[13] However, perhaps because of the importance attributed to conversation in the period, as both a social practice and ethical ideal, contemporaries were particularly inclined to comment on the sonic aspects of the spaces in which they lived and travelled. This book contributes to this pioneering scholarship on sensory history by exploring the integrative quality of hearing and listening, and by showing how the perception of *both* sound *and* noise mediated publicity in rural parishes, urban streets, coffee houses, law courts and colonial spaces of cross-cultural encounter.

The essays in this collection examine the critical role played by judgements about noise and sound in framing the meanings of civility in British discourse and literature in the long eighteenth century. By assessing the impact of shifting understandings of noise and sound, the contributors to this book aim to recover the sonic dimensions of the discourse of civility. With a focus on both the performance and perception of civility in a wide range of contexts, including popular protests and political trials, attacks on Methodism and accounts of London, the book pays particular attention to the way in which censured behaviours and recommended practices resonated beyond the written word. Individually and cumulatively, the essays identify the emergence of civility as a development which radically altered sonic attitudes and experiences, producing new notions of what counted as desirable or undesirable sound. As our contributors show, such changing valuations of noise and sound reveal how contemporaries understood civility during an era of significant political, social and cultural change. More specifically, the book demonstrates that new notions regarding the propriety or impropriety of sound were connected to debates about the impact of urbanisation, the expansion of empire, the rise of popular radicalism, the development of Evangelical religion and the consolidation of legal procedures.

Recently, the magisterial study by Schwartz has been accompanied by several surveys of the history of noise focussed mainly on Western Europe and North America. These surveys have all identified the long eighteenth century as something of a watershed, when experiences and interpretations of noise underwent a thoroughgoing transformation, notably in London. During this period in London, remarks Mike Goldsmith, discrete 'individual noises,' from street cries to animal utterances, were replaced by an indistinct, continuous form of ambient noise, which subsequently became the major source of auditory nuisance.[14] Accordingly, as London became a global city, a centre of trade and empire, it also came to be heard as the noisiest city in the world. Making a similar argument, Garret Keizer has claimed that few sounds in early modern society were oppressive, since they were predominantly 'organic.' By contrast, eighteenth-century London witnessed the emergence of an unvarying flatline noise combined with a perceived rise in the intensity, frequency and irritability of plebeian noises, as heard by the affluent classes.[15] With a shift in focus from London to Britain and its empire, David Hendy has likewise noted that a range of developments, including social and geographical segregation, political revolution, industrialisation, colonialism and the transvaluation of privacy, coalesced to alter the sonic properties and meanings of myriad practices, spaces and social groups. One consequence, among elite auditors, was a new sensitivity and hostility to noise, along with a desire to eradicate, quarantine or withdraw from it.[16] Common to all these studies is the assumption that, beginning in the Enlightenment and climaxing in the Victorian period, noise was increasingly perceived as an unequivocally negative phenomenon.

There is no doubt that the volume of human-produced and instrumental noise swelled substantially in this period under the pressures of urbanisation, industrialisation and unprecedented population growth. Moreover, such developments clearly generated considerable anxieties regarding the effects of noise on the physical health, morality and intellectual improvement of individuals and communities. However, despite this increasing volume, noise was not always perceived as injurious. By listening to the broader range of positive meanings attaching to the perception of noise, our contributors show that far from being necessarily deleterious to civil life, noise was heard in all sorts of positive ways. Across a range of domains, contemporaries found noise useful for understanding states otherwise grasped as abstractions: security, lawfulness, progress, liberty, faith, urbanity, industriousness and so on. Due to the influence of sensibility, for instance, religious worship came to be both policed and celebrated for its association with loud singing and extemporaneous preaching,[17] while liberty was linked to the 'hissing' crowd and progress was registered in acoustic terms as a 'buzz' of activity, well exemplified by coffee-house conversation. Positive meanings of noise seem to have been particularly present among marginalised social groups. As has been shown in relation to the subcultures of pirates, sailors and shoemakers, for example, noise could be associated with qualities such as egalitarianism, independence and a distinctively plebeian style of sociability.[18] At the same time, the French Revolution witnessed the creation of a 'polyvocal public sphere,' which aimed to differentiate between noise and legitimate speech as part of an attempt to balance the 'right to talk' against the 'right to be heard.'[19] Because noise could mark the fluctuating threshold between politeness and vulgarity, a study of the meanings attributed to this acoustic phenomenon makes it possible to recover a wide range of types of civility not yet fully considered, as well as to identify hitherto neglected forms of counter-civility.[20]

By doing so, this book builds on recent scholarship that has established sound studies as a new and exciting field of study, while taking this scholarship into new areas. Our contributors provide an interdisciplinary study of sound and civility in British intellectual, cultural and literary history in the long eighteenth century, a period in which changing forms of civility produced a wide range of new sonic meanings, experiences and values. Literary and cultural histories of noise in twentieth-century Europe and America have proliferated in recent years, significantly advancing the field of sound studies.[21] Similarly, pioneering studies by Bruce Smith and Emily Cockayne have examined the reception and evaluation of noise in sixteenth- and seventeenth-century England; Richard Cullen Rath has written a fascinating reconstruction of sonic meanings in early America; John Picker has produced a remarkable study of soundscapes in the Victorian era; and Mark M. Smith has written a trailblazing acoustic history of nineteenth-century America.[22] However, more work on the social experiences and cultural perceptions of sound in the long eighteenth century in Britain remains to be done. Several excellent edited collections, for example, have examined changing accounts of sound and noise in local, national and global contexts across a number of periods but neglected this crucial moment in the transition to modernity.[23] The essays in this collection chart this neglected terrain, adopting sonic approaches to a range of topics and, in doing so, revealing new connections as well as discontinuities between the Enlightenment, Romantic and Victorian periods.

Historical sound studies have attended to the ways that noise is necessarily a partial, contested and highly charged phenomenon. Being a subjective category, there is usually notable disagreement about what counts as noise, and the interpretation of sound as noise is almost always an expression of social or political conflict, or a disavowal of social or political agency, or both. As Jacques Attali argued in an early elucidation of the topic, historical change is generally accompanied by a heightened sense of noise as a problem, along with a protracted struggle over its meanings.[24] Once again, though, despite this attention to the complexity of noise, historians of sound have largely accepted that it is experienced negatively by affected auditors. Indeed, it is this received understanding of noise as a disruptive surplus that has come to dominate accounts of sonic experience as well as giving shape to a range of legal remedies and political responses to the sounds of urban life.[25]

Perhaps the scholarly consensus regarding the definition of noise as unwanted sound is a product of the omission of the long eighteenth century, a crucial era in the transition to modernity when attitudes to noise were in a state of flux. Our focus on the acoustic aspects of British culture during this period certainly identifies many new contexts, from the conversation piece to the urban riot, in which noise came to be perceived as a problem, an undesirable sound to be regulated, refined or repressed. At the same time, however, several essays in this book also question the apparent self-evidence of noise as an adverse phenomenon. Our book thus recovers a varied soundscape of civility and civic engagement in the period, exploring the making and perceiving of noise across selected media, locations and institutions. Many of the sounds heard in this period were unprecedented yet found a place in the sensorium; for example, regional and national acoustic environments were thoroughly refashioned in response to new forms of social organisation, recreation, political protest, religious belief, mass communication and much else. Importantly, too, the experience of empire infiltrated many aspects of British culture, and it allowed British ideas of civility to be transmitted across the globe. Colonial perceptions of sound and noise have been skilfully and systematically explored in relation to early America, but much work remains to be done on other theatres of empire.[26] Our book will present significant new studies of the sonic dimensions of colonial encounters and colonial societies in the long eighteenth century, focussing on Australia and the Pacific.

The book is structured into four sections, each exploring a different aspect of the relationship between sound and civility. In Part I: *Sound, Conversation and Civility*, two chapters examine the sonic dimensions of conversation, a social practice which was conceived as crucial to the elaboration of civility as well as the production of knowledge. Understandings of civility during the Enlightenment were deeply indebted to the theories of knowledge propounded by John Locke, a philosopher who has often been noted for his privileging of vision over the other senses. As Richard Yeo shows, however, Locke attributed enormous value to sound. Specifically, sound was regarded by Locke as the medium of civil conversation, which for him was the fundamental intellectual touchstone in a turbulent world. John Barrell amplifies the analysis of the normative scope of civil conversation in the eighteenth century in his chapter on the uses and meanings of the 'conversation piece' in British art. For Barrell, a remarkable feature of the conversation piece was that it tended to represent polite people not actually speaking. This was not only the result of the conventions of group portraiture but a reflection of the countless rules governing conversation, which made the performance of civility such a difficult, anxious and contested practice. In each chapter, conversation emerges as the epitome of civil sound – an audible benchmark of cultivated, polite expression in contrast to the rowdy or turbulent noises of less civil engagements.

While civility has conventionally been associated with the purposive sounds of polite conversation and refined conduct, as several essays in this book demonstrate, the audible range of civility throughout this period was surprisingly broad. Shifting attention from theories and representations of conversation to the spaces in which many forms of civility were performed, Part II: *Sonic Spaces of Civility and Incivility* reveals that the sonic range of civility was not restricted to the polite intonation of civil conversation. The chapters in this section explore the reception of both turbulent noise and resonant sound from the early modern period through to the mid-nineteenth century, traversing rural as well as urban, private as well as public, spaces. Focussing on the intersection between sound, gender and status, Dolly MacKinnon provides a groundbreaking study of the sounds of reverberating bells in early modern England, showing how bells rendered gender and social distinctions audible, through tolling as well as the absence of tolling. Markman Ellis examines the new soundscapes of eighteenth-century London and locates within the noisy bustle of coffee houses and the Royal Exchange the discourses of politeness and commercial virtue increasingly associated with the merchant and speculator. In London as well as other spaces, noise was an inevitable if not always desirable accompaniment to the performance of polite social identity, often threatening to blur the lines between civility and incivility. In a wide-ranging analysis, David Ellison considers the Georgian 'shew' house as an ideal location from which to attend to the divergent expression of civil noise. During the late eighteenth and early nineteenth centuries, raucous middle-class visitors to country houses took liberties – in every sense – within spaces which were purpose-built to produce reverberant awe. This mixing of civil and uncivil sound signatures gave rise to new and surprisingly muted forms of sonic differentiation, not least in architecture, registering reverence and power.

Each of the essays in Part III: *Sound, Noise and the Incivility of the Crowd* further complicate our understanding of the sonic range of civility in the late eighteenth and early nineteenth centuries by exploring the audible implications of incivility, especially in relation to popular culture. Focussing on the polite condemnation of popular religious enthusiasm, Peter Denney shows how anti-Methodist writings depicted noise as a key cause of this psychological and social malady, functioning as a dangerous incitement to religious zeal as well as a powerful medium of emotional contagion. Tones of voice, styles of music and modes of utterance were all regarded as important vehicles for spreading religious enthusiasm, and an exposure to instances of acoustic intensity was believed to make people replicate the passionate behaviour of those around them. Turning from religion to popular radicalism, Gillian Russell analyses the collective voice in social and political protest in London in the era of the French Revolution, with a focus on the outdoor meetings of the London Corresponding Society and the alleged attack on George III's coach in late 1795. In her innovative study, Russell argues that 'violent and tumultuous' applause and laughter, along with hissing and hooting, registered popular assertions of rights relating to changing ideas and practices of public life. The audible demarcations of civility and incivility are the subject of Karen Crawley's examination of the transcripts and press coverage of the trials of William Hone, the great radical author and publisher. Each of these three trials, which took place in 1817, reveal the multiple resonances of courtroom sounds: in the rhythm of Hone's speech, the cheers, applause and laughter of the crowd and the judicial appeals to the dispassionate, grave and *quiet* practice of legal judgement. This chapter identifies how sound worked in the courtroom both to constitute and to subvert legal authority, in a comparable manner to the way in which, in relation to popular politics more generally, it could defuse as well as incite rebellious behaviour. Turning to representations of the eighteenth-century crowd in nineteenth-century fiction, Helen Groth analyses the significance of voice and sound in Charles Dickens's treatment of the Gordon Riots in *Barnaby Rudge*. As Groth argues, scenes of riot in this text illuminate a strategic elision of reading and listening through an emphasis on multivocality that aims to affirm the capacity of the novel to function as an archive of ephemeral, individualised acoustic traces. In this way, Victorian fiction trained readers to listen to certain sounds and voices, even as it also screened out a wide range of popular sonic expressions in a highly selective auditory construction of civility for a new audience.

Finally, in Part IV: *Civil and Uncivil Sounds of Empire*, our contributors are among the first scholars to use sound and noise to explore British colonial anxieties about the distinction between civility and incivility. According to Bruce Buchan, civility was a performative virtue, to be seen as well as heard in purposive sounds, the cadences of conversation, of oratory or even of music, each transcending the mere noise of nature or the howling and babbling of the supposedly 'uncivilised.' The experience of empire proved both a confirmation of and a challenge to the self-assurance of European civility, prompting a deep ambivalence about European superiority that was all too often assuaged in the recourse to mere noise. By exploring an important civil case in the penal colony of New South Wales in 1795, Penny Russell hears the echo of vituperative exchanges that rang through the streets, public places and contact zones of early colonial Sydney. According to Russell, insults are particularly revealing colonial voices, and the manner in which they are spoken, heard, misheard and ignored, together with the proximity of speaker and auditor, enriches our understanding not only of the politics of colonial society but also of the role of sound in creating a sense of geography and neighbourliness. In his 'Afterword' to our collection, Mark M. Smith reflects on the place of sound and noise, hearing and listening, in the field of historical sensory studies. Together, the essays in this collection chart the role of under-explored perceptions of and judgements about sound and noise in shaping the understanding of both civil and uncivil activity in the eighteenth and early nineteenth centuries. As our contributors show, noise-makers, hearers and listeners across this dynamic period of European and global history were involved in not only listening to civility *through* but also listening to civility *in* the amplifying noises of a rapidly modernising world.

Notes

1 Lord Shaftesbury, 'The Moralists, A Philosophical Rhapsody' (1709), in *Characteristicks of Men, Manners, Opinions, Times*, ed. Douglas Den Uyl, 3 vols. (1711; Indianapolis: Liberty Fund, 2001), 2:225.
2 Hillel Schwartz, *Making Noise: From Babel to the Big Bang and Beyond* (New York: Zone Books, 2011), 20.
3 David Howes and Constance Classen, *Ways of Sensing: Understanding the Senses in Society* (Abingdon: Routledge, 2014), 110.
4 Schwartz, *Making Noise*, 86–91.

5 Conversation and civility had been closely linked since the sixteenth century, as noted by Anna Bryson, *From Courtesy to Civility: Changing Codes of Conduct in Early Modern England* (Oxford: Oxford University Press, 1998), 160–61. See also Peter Burke, *The Art of Conversation* (Cambridge: Polity, 1993), 89–122. For the continuing connection between conversation and civility in the eighteenth century in relation to associated ideas of politeness, see Philip Carter, *Men and the Emergence of Polite Society* (Harlow: Longman, 2001), 36–37, 63–69; Jon Mee, *Conversable Worlds: Literature, Contention, and Community, 1762–1830* (Oxford: Oxford University Press, 2011), 37–80.

6 Emily Cockayne, *Hubbub. Filth, Noise, and Stench in England, 1600–1770* (New Haven, CT: Yale University Press, 2007), 122.

7 Daniel Defoe, *An Essay Upon Projects* (London: R. Cockerill, 1697), 233–34, 249.

8 Peter Denney, '"The Sounds of Population Fail": Changing Perceptions of Rural Poverty and Plebeian Noise in Eighteenth-Century Britain,' in *Experiences of Poverty in Late Medieval and Early Modern England and France*, ed. Anne M. Scott (Farnham: Ashgate, 2012), 299–301.

9 Jon Stobart, Andrew Hann and Victoria Morgan, *Spaces of Consumption: Leisure and Shopping in the English Town, c.1680–1830* (Abingdon: Routledge, 2007), 86–87; Mark M. Smith, *Sensing the Past: Seeing, Hearing, Smelling, Tasting, and Touching in History* (Berkeley: University of California Press, 2007), 45.

10 Cockayne, *Hubbub*, 112–113.

11 Peter Bailey, 'Breaking the Sound Barrier: A Historian Listens to Noise,' *Body and Society* 2, no. 2 (1996): 50.

12 Alain Corbin, *Village Bells: Sound and Meaning in the Nineteenth-Century French Countryside*, trans. Martin Thom (New York, NY: Columbia University Press, 1998), 96.

13 William Tullett, 'The Macaroni's "Ambrosial Essences": Perfume, Identity and Public Space in Eighteenth-Century England,' *Journal for Eighteenth-Century Studies* 38, no. 2 (2015): 163–80; Marlene Eberhardt, 'Sensing Space and Making Publics,' in *Making Space Public in Early Modern Europe: Geography, Performance, Privacy*, eds. Angela Vanhaelen and Joseph P. Ward (Abingdon: Routledge, 2013), 173–89; Peter Denney, 'Looking Back, Groping Forward: Rethinking Sensory History,' *Rethinking History* 15, no. 4 (2011): 601–4.

14 Mike Goldsmith, *Discord: The Story of Noise* (Oxford: Oxford University Press, 2012), 77.

15 Garret Keizer, *The Unwanted Sound of Everything We Want: A Book about Noise* (New York: Public Affairs, 2010), 94–98.

16 David Hendy, *Noise: A Human History of Sound and Listening* (London: Profile Books, 2013), 167–210.

17 Phyllis Mack, 'The Senses in Religion: Listening to God in the Eighteenth Century,' in *A Cultural History of the Senses in the Age of Enlightenment*, ed. Anne C. Vila (London: Bloomsbury, 2014), 85–107. See also Leigh Eric Schmidt, *Hearing Things: Religion, Illusion and the American Enlightenment* (Cambridge, MA: Harvard University, 2000), 38–77.

18 For the significance of noise among shoemakers, see Peter Denney: 'Clamoring for Liberty: Alehouse Noise and the Political Shoemaker,' *Eighteenth-Century Life* 41, no. 2 (2017): 105–21; for the role of noise in the cultures of pirates and sailors, see Bruce Buchan, 'Civility at Sea: From Murmuring to Mutiny,' *Republics of Letters* 5, no. 2 (2017): 1–14.

19 Sophie Rosenfeld, 'On Being Heard: A Case for Paying Attention to the Historical Ear,' *American Historical Review* 116, no. 2 (2011): 327–34.

20 See David Ellison, 'The Call Came from Inside the House: Supernatural Noise and Domestic Incivility,' *Republics of Letters* 5, no. 2 (2017): 1–12.
21 See, for example, James Mansell, *The Age of Noise in Britain: Hearing Modernity* (Urbana: University of Illinois Press, 2017); Sam Halliday, *Sonic Modernity: Representing Sound in Literature, Culture and the Arts* (Edinburgh: Edinburgh University Press, 2013); Karen Bijsterveld, *Mechanical Sound: Technology, Culture and Public Problems of Noise in the Twentieth Century* (Cambridge, MA: MIT Press, 2008); Emily Thompson, *The Soundscape of Modernity: Architectural Acoustics and the Culture of Listening in America, 1900–1933* (Cambridge, MA: MIT Press, 2002).
22 Bruce R. Smith, *The Acoustic World of Early Modern England: Attending to the O-Factor* (Chicago: University of Chicago Press, 1999); Cockayne, *Hubbub*; Richard Cullen Rath, *How Early America Sounded* (Ithaca, NY: Cornell University Press, 2003); John M. Picker, *Victorian Soundscapes* (New York, NY: Oxford University Press, 2003); Mark M. Smith, *Listening to Nineteenth-Century America* (Chapel Hill: University of North Carolina Press, 2001).
23 See, for example, Ian Biddle and Kirsten Gibson, eds., *Cultural Histories of Noise, Sound and Listening in Europe, 1300–1918* (Abingdon: Routledge, 2016); Daniel Morat, ed., *Sounds of Modern History: Auditory Cultures in Nineteenth- and Twentieth-Century Europe* (New York: Berghahn, 2014); Joy Damousi and Desley Deacon, eds., *Talking and Listening in the Age of Modernity: Essays on the History of Sound* (Canberra: ANU E-Press, 2007); Veit Erlmann, ed., *Hearing Cultures: Essays on Sound, Listening and Modernity* (Oxford: Berg, 2004).
24 Jacques Attali, *Noise: The Political Economy of Music,* trans. Brian Massumi (Minneapolis: University of Minnesota Press, 1985).
25 Bijsterveld, *Mechanical Sound*; Picker, *Victorian Soundscapes.*
26 Rath, *How Early America Sounded.* See also Diane Collins, 'Acoustic Journeys: Exploration and the Search for an Aural History of Australia,' *Australian Historical Studies* 37, no. 128 (2006): 1–17; Andrew J. Rotter, 'Empire of the Senses: How Seeing, Hearing, Smelling, Tasting, and Touching Shaped Imperial Encounters,' *Diplomatic History* 35, no. 1 (2011): 11–12; Joy Damousi, *Colonial Voices: A Cultural History of English in Australia, 1840–1940* (Cambridge: Cambridge University Press, 2010).

Bibliography

Primary sources

Defoe, Daniel. *An Essay Upon Projects.* London: R. Cockerill, 1697.
Shaftesbury, Lord. *Characteristicks of Men, Manners, Opinions, Times.* 1711. Edited by Douglas Den Uyl, 3 vols. Indianapolis, IN: Liberty Fund, 2001.

Secondary sources

Attali, Jacques. *Noise: The Political Economy of Music.* Translated by Brian Massumi. Minneapolis: University of Minnesota Press, 1985.
Bailey, Peter. 'Breaking the Sound Barrier: A Historian Listens to Noise.' *Body and Society* 2, no. 2 (1996): 49–66.
Biddle, Ian and Kirsten Gibson, eds. *Cultural Histories of Noise, Sound and Listening in Europe, 1300–1918.* Abingdon: Routledge, 2016.

Bijsterveld, Karen. *Mechanical Sound: Technology, Culture and Public Problems of Noise in the Twentieth Century.* Cambridge, MA: MIT Press, 2008.
Bryson, Anna. *From Courtesy to Civility: Changing Codes of Conduct in Early Modern England.* Oxford: Oxford University Press, 1998.
Buchan, Bruce. 'Civility at Sea: From Murmuring to Mutiny.' *Republics of Letters* 5, no. 2 (2017): 1–14.
Burke, Peter. *The Art of Conversation.* Cambridge: Polity, 1993.
Carter, Philip. *Men and the Emergence of Polite Society.* Harlow: Longman, 2001.
Cockayne, Emily. *Hubbub. Filth, Noise, and Stench in England, 1600–1770.* New Haven, CT: Yale University Press, 2007.
Collins, Diane. 'Acoustic Journeys: Exploration and the Search for an Aural History of Australia.' *Australian Historical Studies* 37, no. 128 (2006): 1–17.
Corbin, Alain. *Village Bells: Sound and Meaning in the Nineteenth-Century French Countryside.* Translated by Martin Thom. New York: Columbia University Press, 1998.
Damousi, Joy. *Colonial Voices: A Cultural History of English in Australia, 1840–1940.* Cambridge: Cambridge University Press, 2010.
Damousi, Joy and Deacon, Desley, eds. *Talking and Listening in the Age of Modernity: Essays on the History of Sound.* Canberra: ANU E-Press, 2007.
Denney, Peter. 'Clamoring for Liberty: Alehouse Noise and the Political Shoemaker.' *Eighteenth-Century Life* 41, no. 2 (2017): 105–121.
Denney, Peter. 'Looking Back, Groping Forward: Rethinking Sensory History.' *Rethinking History* 15, no. 4 (2011): 601–616.
Denney, Peter. '"The Sounds of Population Fail": Changing Perceptions of Rural Poverty and Plebeian Noise in Eighteenth-Century Britain.' In *Experiences of Poverty in Late Medieval and Early Modern England and France*, edited by Anne M. Scott, 295–311. Farnham: Ashgate, 2012.
Eberhardt, Marlene. 'Sensing Space and Making Publics.' In *Making Space Public in Early Modern Europe: Geography, Performance, Privacy*, edited by Angela Vanhaelen and Joseph P. Ward, 173–89. Abingdon: Routledge, 2013.
Ellison, David. 'The Call Came from Inside the House: Supernatural Noise and Domestic Incivility.' *Republics of Letters* 5, no. 2 (2017): 1–12.
Erlmann, Veit, ed. *Hearing Cultures: Essays on Sound, Listening and Modernity.* Oxford: Berg, 2004.
Goldsmith, Mike. *Discord: The Story of Noise.* Oxford: Oxford University Press, 2012.
Halliday, Sam. *Sonic Modernity: Representing Sound in Literature, Culture and the Arts.* Edinburgh: Edinburgh University Press, 2013.
Hendy, David. *Noise: A Human History of Sound and Listening.* London: Profile Books, 2013.
Howes, David and Constance Classen. *Ways of Sensing: Understanding the Senses in Society.* Abingdon: Routledge, 2014.
Keizer, Garret. *The Unwanted Sound of Everything We Want: A Book About Noise.* New York: Public Affairs, 2010.
Mack, Phyllis. 'The Senses in Religion: Listening to God in the Eighteenth Century.' In *A Cultural History of the Senses in the Age of Enlightenment*, edited by Anne C. Vila, 85–107. London: Bloomsbury, 2014.

Mansell, James. *The Age of Noise in Britain: Hearing Modernity*. Urbana: University of Illinois Press, 2017.

Mee, Jon. *Conversable Worlds: Literature, Contention, and Community, 1762–1830*. Oxford: Oxford University Press, 2011.

Morat, Daniel, ed. *Sounds of Modern History: Auditory Cultures in Nineteenth- and Twentieth-Century Europe*. New York: Berghahn, 2014.

Picker, John M. *Victorian Soundscapes*. New York: Oxford University Press, 2003.

Rath, Richard Cullen. *How Early America Sounded*. Ithaca, NY: Cornell University Press, 2003.

Rosenfeld, Sophie. 'On Being Heard: A Case for Paying Attention to the Historical Ear.' *American Historical Review* 116, no. 2 (2011): 316–34.

Rotter, Andrew J. 'Empire of the Senses: How Seeing, Hearing, Smelling, Tasting, and Touching Shaped Imperial Encounters.' *Diplomatic History* 35, no. 1 (2011): 3–19.

Schmidt, Leigh Eric. *Hearing Things: Religion, Illusion and the American Enlightenment*. Cambridge, MA: Harvard University Press, 2000.

Schwartz, Hillel. *Making Noise: From Babel to the Big Bang and Beyond*. New York: Zone Books, 2011.

Smith, Bruce R. *The Acoustic World of Early Modern England: Attending to the O-Factor*. Chicago: University of Chicago Press, 1999.

Smith, Mark M. *Listening to Nineteenth-Century America*. Chapel Hill: University of North Carolina Press, 2001.

Smith, Mark M. *Sensing the Past: Seeing, Hearing, Smelling, Tasting, and Touching in History*. Berkeley: University of California Press, 2007.

Stobart, Jon, Andrew Hann, and Victoria Morgan. *Spaces of Consumption: Leisure and Shopping in the English Town, c.1680–1830*. Abingdon: Routledge, 2007.

Thompson, Emily. *The Soundscape of Modernity: Architectural Acoustics and the Culture of Listening in America, 1900–1933*. Cambridge, MA: MIT Press, 2002.

Tullett, William. 'The Macaroni's "Ambrosial Essences": Perfume, Identity and Public Space in Eighteenth-Century England.' *Journal for Eighteenth-Century Studies* 38, no. 2 (2015): 163–80.

Part I

Sound, conversation and civility

1 John Locke on sound and conversation

Richard Yeo

> Pleasures invade both Eye and Ear,
> So fierce the transports are, they wound,
> And all my senses feasted are,
> Tho' yet the Treat is only Sound.[1]

John Locke (1632–1704) was an English physician and Fellow of the Royal Society of London. He became famous as the author of the *Essay Concerning Human Understanding* (1690) and *Two Treatises of Government* (1690), and henceforth was regarded as a 'philosopher' and the reputed founder of English 'empiricism,' a term he never used. In this chapter, I suggest that by recognising that Locke was not only a philosopher, but also a physician, traveller and scientific *virtuoso*, we may deepen our understanding of his attitude to sound, in these ways – as one of the five senses; as a negative presence, in 'noise'; and as the medium of civil conversation.

In 1656, the young Locke witnessed the arraigning of the radical preacher James Nayler (1618?–60) and other dissident Quakers at Westminster Hall, London. He wrote to his father, saying that:

> Nailer one man more and 3 or 4 women of the tribe all with white gloves and the womens heads in white baggs. their carriage was strange to me, one of the women made a continued humming noise longer then the reach of an ordinari breath without motion either of lips or breath that I who stood next her could perceive. [they did not answer questions put to them] or did it with a great deale of suttlety besides the cover and cunning of that language which others and I beleeve they them selves scarce understand.[2]

Although the Quakers were a small, tight sect, Locke came to see them as exemplifying a more general feature of human nature: namely, that individuals bound together by strong sympathies influenced each other, often in deleterious ways. In October 1659, in a letter to a friend, he declared that 'we are all Quakers here and there is not a man but thinks he alone hath this light within and all besids stumble in the darke.'[3] In Locke's estimate, the Quakers were 'enthusiasts' who felt sure in their breasts that they were guided by an inner light direct from God. He added a separate chapter on 'Enthusiasm' to the fourth edition of the *Essay*.[4] For Locke, intelligible conversation relied on spoken words acting as signifiers of '*Ideas*.'[5] The link between word and idea was conventional and hence required a social contract. Such a contract might well be limited to two persons, or to the members of a sect like the Quakers, but this could be dangerous. At the encounter in London in 1656, Locke noted that some human sounds, apparently serving as communication, could appear as mere incoherent mumbling, or noise, to those outside the circle of initiates.

The senses

The question of what Locke thought about 'sound' can be approached by considering his answer, in the *Essay*, to the question of how individuals acquired knowledge. He considered the five senses as the avenues by which information of the world reached the mind. By his account, this took place via a sequence of sensations, perceptions, formation of simple ideas, comparative reflection by the 'Judgment' on these, combinations of such ideas as complex ideas and the retention of ideas in memory.[6] To some extent, this model of mental processing was the Aristotelian one, as elaborated by subsequent thinkers, including the Scholastics. It was taken as a starting point by Locke and his contemporaries, such as Thomas Hobbes and René Descartes.[7]

One consequence of this model was that knowledge produced in the mind depended on input from the senses, which themselves might be weak or faulty. Accordingly, there was fascination with the notion of augmenting the senses – for example, by taking drugs, or by the use of instruments such as spectacles, telescopes and microscopes or, in the case of hearing, by the attachment of various ear trumpets. Another speculation was that the senses could be *bypassed* altogether. John Evelyn, diarist, virtuoso and Fellow of the Royal Society of London, told of someone who contended that 'if a hole could dextrously be boar'd through the Skull to the Brain, in the midst of the Fore-head, a man might both see, & hear, & smell without the use of any other Organs.' This person was 'one Tom Whittal, a student of Christ-Church' (incidentally, Locke's college at Oxford).[8] Yet, however input to the brain was delivered, Locke maintained that the cognitive outcome depended on analysis and reflection conducted by the reason or judgement, often referred to simply as 'the mind.'[9]

In his *Essay*, Locke offered a rigorous account of the *distinctive* input from the various senses. He regarded specific ideas as dependent on information from specific senses:

> But all that are born into the World being surrounded with Bodies, that perpetually and diversely affect them, variety of *Ideas*, whether

> care be taken about it or no, are imprinted on the Minds of Children. *Light*, and *Colours*, are busie at hand every where, when the Eye is but open; *Sounds*, and some *tangible Qualities* fail not to solicite their proper Senses, and force an entrance to the Mind; but yet, I think, it will be granted easily, That if a Child were kept in a place, where he never saw any other but Black and White, till he were a Man, he would have no more *Ideas* of Scarlet or Green, than he that from his Childhood never tasted an Oyster, or a Pine-Apple, has of those particular Relishes.[10]

And a little further on:

> The same inability, will every one find in himself, who shall go about to fashion in his Understanding any simple *Idea*, not received in by his Senses, from external Objects, or by reflection from the Operations of his own mind about them. I would have any one try to fancy any Taste, which had never affected his Palate; or frame the *Idea* of a Scent, he had never smelt: And when he can do this, I will also conclude, that a blind Man hath *Ideas* of Colours, and a deaf Man true distinct Notions of Sounds.[11]

What later became known as the 'Molyneux problem' involved a separation of sensations delivered by two of the senses, sight and touch. On 2 March 1693, Locke's friend, the Irish politician and philosopher William Molyneux (1656–98), wrote from Dublin about a question stimulated by a passage in the *Essay*:

> I will conclude my tedious lines with a Jocose Problem, that, upon Discourse with several concerning your Book and Notions, I have proposed to divers very ingenious Men, … tis this: Suppose a Man born blind, and now adult, and taught by his Touch to Distinguish between a Cube and a Sphere (Suppose) of Ivory, nighly of the same Bignes, so as to tel, when he felt One and tother, which is the Cube which the Sphaere. Suppose then, the Cube and Sphaere placed on a Table, and the Blind man to be made to see. Quaere whether by his sight, before he touchd them, he could now Distinguish and tel which is the Globe which the Cube. I answer, Not; for tho he has obtaind the Experience of How a Globe, how a Cube affects his Touch. Yet he has not yet attaind the Experience, that what affects my Touch so or so, must affect my Sight so or so; or that a Protuberant Angle in the Cube that presd his hand unequally, shall appear to his Eye as it does in the Cube. But of this enough; perhaps you may find some place in your Essay, wherein you may not think it amis to say something of this Problem.[12]

In adding his response to this problem in the fourth edition of the *Essay* (1700), Locke said:

> I agree with this thinking Gent. whom I am proud to call my Friend, in his answer to this his Problem; and am of opinion, that the Blind Man, at first sight, would not be able with certainty to say, which was the Globe, which the Cube, whilst he only saw them: though he could unerringly name them by his touch, and certainly distinguish them by the difference of their Figures felt. This I have set down, and leave with my Reader, as an occasion for him to consider, how much he may be beholding to experience, improvement, and acquired notions, where he thinks he has not the least use of, or help from them.[13]

Locke did not use this thought experiment to show the greater power of one sense over another; rather, his point concerned the inability of one sense, in the *short term* of this scenario, to supply information to another.[14] However, such comparison and ranking of the different senses was common in his day.

The hierarchy of the senses

With many of his contemporaries, Locke often privileged the sense of sight, referring in the *Essay* to 'that most instructive of our Senses, Seeing,' and calling 'Sight, the most comprehensive of all our Senses.'[15] This judgement had a long past. In 1585, Giovanni Battista Benedetti (1530–90), a court mathematician to the Duke of Savoy, summarised Aristotle's view that vision, more than any other sense, leads to knowledge; from this perspective, we think by recalling visual images stored in memory.[16] This opinion is found in Richard Braithwaite's *Essaies upon the five senses* (London 1620), in which the order of the senses, treated in individual chapters, is given as seeing, hearing, touching, tasting and smelling.[17] Well into the seventeenth century, as Stuart Clark has shown in his *Vanities of the Eye* (2007), the dominant judgement was that vision was the premier sense: it was closest to the all-seeing 'watchfulness' of God; for Saint Augustine, it provided an analogue for 'spiritual vision'; and in the long-standing account of how sensations were transformed into knowledge, mental replicas of things in the world were seen by the 'eye' of the mind. (This view was maintained in spite of the increasing doubts about whether sight ensured veridical pictures of the external world.)[18] In these discussions, hearing was usually the main comparison, in keeping with its status as second in the hierarchy of senses. However, there was a minority position that, while acknowledging that vision was the most comprehensive sense, nevertheless affirmed the priority of hearing for what Aristotle called 'rational discourse' and learning.[19] Locke agreed with this, and his comments on the senses of sight and sound often involved a consideration of their role in civil conversation.

Locke speculated about improving vision and hearing by technical means and considered the likely results. He predicted that if an individual's power of 'seeing' were '1000, or 100000 times more acute than it is now by the best Microscope,' then such a person would 'come nearer the Discovery of the Texture and Motion of the minute Parts of corporeal things.' However, he argued that such a 'man' (in Locke's example) with 'Microscopical Eyes' would be ill-suited to this world: he 'would be in a quite different World from other People: Nothing would appear the same to him, and others... So that I doubt, Whether he, and the rest of Men, could discourse concerning the Objects of Sight.' Locke also imagined a scenario in which 'our Sense of Hearing were but 1000 times quicker than it is,' also concluding that the outcome would be negative: 'how would a perpetual noise distract us. And we should in the quietest Retirement, be less able to sleep or meditate, than in the middle of a Sea-fight.'[20] Presumably, this was the loudest noise Locke could imagine.

There is an intriguing *asymmetry* between these two scenarios: the augmentation of vision seems to work, although the person with microscopic sight is unable to convey to others ideas formed with the aid of such magnification. Yet in the case of amplified hearing, the result is almost totally non-functional: this individual perceives only undifferentiated 'noise' – an effect, we might suppose, almost tantamount to deafness. However, the crucial similarity between the two scenarios may be that neither the enhancement of sight nor hearing results in improved communication in the social world. For Locke, this was a fundamental requirement, and this is why he agreed with those who pressed the importance of the sensation of sound as the primary medium of intellectual exchange.

In the *Essay*, there are about six times as many references to being blind than to being deaf. This attests to the primacy of vision in discussions about the generation of knowledge in the mind, but there is no doubt that Locke agreed with Aristotle that hearing was the most crucial sense for the development of knowledge and the ability to converse in the world.[21] After Locke's death in October 1704, the philosopher Damaris Masham responded to a request for her memories of 'our dead Friend.'[22] She reported that Locke had once said that he would rather 'be Blind than Deaf.'[23] Similarly, Michel de Montaigne, in his essay 'On the Art of Conversation,' declared that

> To my taste the most fruitful and most natural exercise of our minds is conversation... That is why, if I were now obliged to make the choice, I think I would rather lose my sight than my powers of speech or hearing.[24]

This was Locke's view and, as a physician interested in the 'art of longevity,' he was on the lookout for the effects of age on the senses. In a journal entry for 1 March 1681, he recorded meeting in Oxford a woman,

Alice George, who was reputedly 108 years old. Among other observations of her condition, Locke noted that 'her hearing is very good.'[25] Locke's estimate of the sense of sound may have been confirmed by an illness during 1702 in which he suffered severe loss of hearing. Towards the end of that year he told his friend, Edward Clarke, that

> I have been little better than out of the world these last twelve months by a deafness that in great measure shut me out of conversation. I thank god my hearing is now restored again and it is in your power to make me yet more sensible of that blesseing.[26]

Conversations

For Locke, civil conversation was the fundamental intellectual touchstone in a turbulent world. Conversation involved a fragile relationship between two or more people. Whatever improvements might be made to the sense of hearing, conversation depended on social conventions, and all people had a moral responsibility to respect and refine these. Sometimes Locke has in mind the older meaning of 'conversation' as the act of living with and sharing the company of others. An English translation of Cervantes's *Don Quixote* in 1620 has it that 'You may know the man, by the conversation he keepes,' referring primarily to the company in which that person lived.[27] Locke also asserted that, in most situations, the language of conversation should be accessible to all reasonable persons. The Quakers failed this test; so, too, did anyone speaking only in the terms of their sect or school.[28] Such uncivil people included some of Locke's teachers at Oxford in the 1650s who carried on the disputatious jargon of neo-scholasticism. This verbose wrangling over poorly defined topics and questions might not matter if it remained within small groups or if such a practice was easily shed by individuals upon leaving the universities.[29] However, Locke announced that the disease had spread: 'Nor hath this mischief stopped in logical Niceties, or curious empty Speculations; it hath invaded the great Concernments of Humane Life and Society.'[30] It is striking that when Locke came to describe pain and pleasure, this kind of improper conversation featured in his operational definition: 'The pain of tender Eyes, and the pleasure of Musick; Pain from captious uninstructive wrangling, and the pleasure of rational conversation with a Friend, or of well directed study in the search and discovery of Truth.'[31]

Locke belonged to several small conversation groups.[32] These included Robert Boyle's experimental circle in Oxford from about 1660; an informal group of friends meeting in his room at Exeter House, the London residence of Anthony Ashley Cooper (Earl of Shaftesbury from 1672), which is probably the group referred to in the 'The Epistle to the Reader' of the *Essay*; a *salon* hosted by the scholar Henri Justel (1620–93) in

Paris in the mid-1670s; a philosophical discussion group in Amsterdam involving Jean Le Clerc and Philip van Limborch; a possible successor to this in Rotterdam from late 1686, meeting in Benjamin Furly's house; and finally, the 'Dry Club' in London (possibly from 1689, or not until 1692) after Locke's return from the Netherlands on 12 February 1689, just as the Glorious Revolution was inaugurated, with William and Mary of Orange jointly accepting the English throne.[33] Pierre Des Maizeaux, in *A Collection of Pieces* (1720), wrote that 'Mr. Locke took a delight in forming such Societies, wherever he made any stay.'[34] He mentioned one in Amsterdam and a continuation of this in London, and he published a set of 'Rules for a Society,' attributing these to Locke, although there is no surviving manuscript copy among Locke's papers.[35] However, it is very likely that Locke did draw up a set of rules to govern the discussion of the 'Dry Club.' A manuscript copy, dated 1692, and titled 'The Rules of the Dry Club: For the Amicable Improvement of Mix'd Conversation,' is among his extant papers.[36]

There are significant resemblances between the membership rules (as opposed to other details) of the Dry Club and those Des Maizeaux published.[37] Entry to these conversation groups required nomination by an existing member who vouched that the nominee respected all men irrespective 'of what Church or Profession of Religion soever they are.'[38] This condition resonated with Locke's interest in religious toleration, a topic on which he was writing in the late 1680s. He published *Epistola de Tolerantia* in 1689; this was quickly translated into English by William Popple.[39] Once membership was settled, the remaining rules aimed at ensuring timely and disciplined discussion of agreed topics under the direction of a 'Proposer' who ensured that members spoke, uninterrupted, in turn. One of the rules of the Dry Club was specific about the kind of talk expected: 'Every one endeavouring that their Answers may not be onely loose Discourses upon the Subject; but that they tend directly to, and (as much as possible) end in the Resolution of the Question proposed.' Another prescribed that no member should 'mention the Name either of any Person liveing, present or absent, or of any Church, Sect, or Society of Men whatsoever, in this Island.'[40] Locke sought to avoid contentious party and sectarian disputes, and his rules allowed that 'any appearance of growing Warmth' may be the moment to terminate a discussion.[41]

The historian Philip Milton has remarked that the Dry Club rules may seem 'unduly complex for a group of friends meeting for discussion,' but he also notes that they display 'a fussy attention to detail that is entirely characteristic' of Locke.[42] We have evidence that some friends believed that certain strictures were too severe: Benjamin Furly (1636–1714), an English Quaker merchant living in Rotterdam, thought it unnecessary to demand that 'every one speak to every question'; this was onerous, especially for new members, and he told Locke that it may be 'enough to terrify a modest man from offering himself to your society.'[43] One of Locke's biographers has suggested that he preferred small conversation groups, able to be controlled, in contrast to the larger and more robust gatherings of the Royal Society, whose meetings he seldom attended.[44] This may be so, and indeed, there is a further possible refinement: namely, that most of Locke's untrammelled conversations did not take place in the surroundings governed by his own rules, but in more intimate, personal encounters. This is implicit in his remark, reported by Damaris Masham, that among the 'contentments' he enjoyed in London after his return from The Netherlands in 1689 was 'that of spending one day every week with my lord Pembroke in a Conversation undisturb'd by such as could not beare a part in the best entertainment of Rational Minds, Free Discourse concerning usefull Truths.'[45] However, we should not conclude from this strong hint of exclusivity that Locke found no place for other types of talk: with children, with strangers and, admitting a proper mixture of 'Raillery,' with friends.[46]

As well as thinking about conventions that could monitor behaviour likely to compromise civil conversation, Locke recognised that there were threats from within, so to speak. In one of his last works, *Conduct of the Understanding* (1697), he confronted the ways in which the passions could affect proper thinking and the conversations that shaped it and conveyed it to others. He put the general concern thus:

> Matters, that are recommended to our thoughts by any of our passions, take possession of our minds with a kind of authority, and will not be kept out or dislodged; but, as if the passion that rules were, for the time, the sheriff of the place, and came with all the posse, the understanding is seized and taken with the object it introduces, as if it had a legal right to be alone considered there. There is scarce any body, I think, of so calm a temper who hath not some time found this tyranny on his understanding, and suffered under the inconvenience of it.[47]

Of individuals so afflicted, he wrote that 'They see not what passes before their eyes; hear not the audible discourse of the company.' In other words, their two primary senses were debilitated. Locke listed other disturbances, such as obsessive thoughts, recurring 'visions' and annoying sounds (or 'earworms,' in today's terms), such as 'some trivial sentence, or a scrap of poetry,' and most alarmingly, 'the chiming of some particular words or sentence in the memory, and, as it were, making a noise in the head, and the like,... [or even] the insignificant buzz of purely empty sounds.'[48] One remedy, he suggested, was to focus the mind on something worthwhile, so as to 'set it free from these idle companions,' from these intrusive thoughts and sounds.[49] Surely, however, from what we know of Locke's own analyses and preferences, another solution was to talk with a suitable companion, a rational friend, about all manner of things.

As with all knowledge of the world, Locke tracked '*Ideas*' back to their sensory sources. In the case of sound, the sense of hearing was the relevant channel, one that could deliver a range of perceptions from noise to music to intelligible communication. The latter, in the form of conversation, could be enlivening and enriching, but when individuals were confined within misguided sects (perhaps, the Quakers) or swayed by immoral clubs, even conversation could endanger the prospect of eternal happiness. Locke's close analysis of sound as a sensory input and source of ideas about the world was matched by his vigilance regarding conversation. The conversations that individuals had with themselves, and their discussions in the social world, had to be evaluated. Despite these demanding injunctions, Locke believed that communication among people was a major source of both knowledge and pleasure.

Acknowledgement

I thank Bruce Buchan and Peter Denney for their helpful comments on this chapter. For permission to consult manuscript material I thank The British Library, London and the Bodleian Library, Oxford.

Notes

1 'A Song set by Mr. Henry Purcell, the Words by Coloniel Heveningham,' *Gentleman's Journal, or, The Monthly Miscellany* (1692): 7–29.
2 Locke to John Locke, senior, 15 November 1656, in *The Correspondence of John Locke*, ed. E.S. de Beer, 8 vols. (Oxford: Clarendon Press, 1976–1989), 1:44.
3 Locke to Tom [Thomas Westrowe(?)], 20 October 1659, in *Correspondence*, 1:123.
4 John Locke, *An Essay Concerning Human Understanding* (1690), ed. Peter H. Nidditch (Oxford: Clarendon Press, 1975; incorporating 4th edn of 1700), 4:xix.
5 In the *Essay*, Locke invariably put '*Idea*s' in italics. I follow this only when quoting from him.
6 Locke, *Essay*, 2:1–3 on simple ideas; 4:xiv on 'Judgment.'
7 Ruth E. Harvey, *The Inward Wits: Psychological Theory in the Middle Ages and the Renaissance* (London: Warburg Institute, 1975).
8 John Evelyn to John Beale, 27 August 1668, British Library, Add MS 78298, fols 172^{v}–174^{r}.
9 Locke, *Essay*, 1.ii.
10 Locke, *Essay*, 2.i.6.
11 Locke, *Essay*, 2.ii.2.
12 William Molyneux to Locke, 2 March 1693, *Correspondence*, 4:651.
13 Locke, *Essay*, 2.ix.8.
14 For recent experimental evidence supporting the conclusion of Locke and Molyneux, see R. Held et al., 'The Newly Sighted Fail to Match Seen with Felt,' *Nature Neuroscience* 14, no. 5 (2011): 551–53.

15 Locke, *Essay*, 2.xxiii.12; 2.ix.9.

16 Giovanni Battista Benedetti, '*Comparitio Visus, et Auditus*,' in his *Diversarum Mathematicarum et Physicarum Liber* (Turin, 1585), 414–16, cited in Thomas Frangenberg, '*Auditus Visu Prestantior*: Comparisons of Hearing and Vision in Charles de Bovelles's *Liber de Sensibus*,' in *The Second Sense: Studies in Hearing and Musical Judgement from Antiquity to the Seventeenth Century*, eds. C. Burnett, M. Fend and P. Gouk (London: Warburg, 1991), 71–74.

17 Richard Braithwaite, *Essaies Upon the Five Senses* (London: E. G[riffin], 1620). Robert Hooke (1635–1703), who, like Locke, was a Fellow of the Royal Society, gave this order of the senses: 'the ist. and most Spiritual is plac'd in the Eye, a 2d. in the Ear, a 3d. in the Nose, a 4th. in the Mouth, the 5th. over the whole Body.' Hooke, 'A General Scheme, or Idea of the Present State of Natural Philosophy' [written in circa 1668], in *The Posthumous Works of Robert Hooke*, ed. Richard Waller (London: Smith and Walford, 1705), 12.

18 Stuart Clark, *Vanities of the Eye*: *Vision in Early Modern European Culture* (Oxford: Oxford University Press, 2007), 9–12.

19 Aristotle, *De Sensu et Sensibili*, 437a4–437a16, in *The Complete Works of Aristotle, the Revised Oxford Translation*, ed. Jonathan Barnes (Princeton, NJ: Princeton University Press, 1984) vol. 1; Clark, *Vanities*, 20–21. For a vigorous assertion of the primacy of hearing, see Charles de Bovelles (1479–1567), *Liber de Sensibus* (Paris, 1511), discussed in Frangenberg, 'Auditus Visu Prestantior,' 74–89.

20 Locke, *Essay*, 2.xxiii.12.

21 In a recent study asking people whether they would prefer to be blind or deaf, 17 per cent chose blindness, many citing the loss of music as the worst consequence of deafness; see Jiye G. Kim, Andrew J. Goldman, and Irving Biederman, 'Blind or Deaf?: A Matter of Aesthetics,' *Perception* 37, no. 6 (2008): 949–50.

22 Damaris Cudworth, the daughter of the Cambridge philosopher Ralph Cudworth (1617–88), married the politician Francis Masham. Locke lived with them in their house ('Oates') in Essex from January 1691 until his death on 28 October 1704.

23 Damaris Masham to Jean Le Clerc, 12 January 1705, in Jean Le Clerc, *Epistolario*, ed. Maria Grazia and Mario Sina, 4 vols. (Firenze: L. S Olschki, 1987–1994), 2:497. The full remark from the final passage of Masham's letter is illegible in the original manuscript held in the Universiteits-Bibliotheek, Amsterdam (MS R. K., J 57 a); Grazia and Sina do not print it. However, it occurs in this form in the English translation of Jean Le Clerc's transcription in his *The Life and Character of Mr John Locke ... done into English by T.F.P. Gent* (London: John Clark, 1706), 30. This work was originally published in the *Bibliothèque Choisie* 6 (1705), 342–411.

24 Michel de Montaigne, 'On the Art of Conversation,' in *The Complete Essays*, trans. M.A. Screech (London: Penguin Books, 1991), Book 3, no. 8, 1045. The *Essais* were published in 1580; this essay appeared in the second edition of 1588.

25 See Locke's journal in the Bodleian Library, Oxford, MS Locke f. 5, 1 March 1681, 19–22, at 20. See also Locke to Nicolas Toinard, 1 April 1681, *Correspondence*, vol. 2, letter no. 633, 394–96, for 'her art of longevity' (translated from the Latin by E.S. De Beer, at p. 396). For records of information, including conversations, in Locke's journals, see Richard Yeo, *Notebooks, English Virtuosi, and Early Modern Science* (Chicago: University of Chicago Press, 2014), 195–201.

26 John Locke to Edward Clarke, 30 November 1702, *Correspondence*, 7:718.
27 Miguel de Cervantes, *The Second Part of the History of ... Don Quixote of the Mancha*, trans. Thomas Shelton (London: E. Blount, 1620), 458. For the nuances of friendship and conversation in early modern contexts, see Keith Thomas, *The Ends of Life: Roads to Fulfilment in Early Modern England* (Oxford: Oxford University Press, 2009), 187–225.
28 Locke, *Essay*, 2.xiii.27.
29 See Richard Yeo, 'John Locke and Polite Philosophy,' in *The Philosopher in Early Modern Europe: The Nature of a Contested Identity*, eds. C. Condren et al. (Cambridge: Cambridge University Press, 2006), 267–68, 270–71.
30 Locke, *Essay*, 3.x.12.
31 Locke, *Essay*, 2.xx.18.
32 For information on the following examples, see Jean Le Clerc, *Bibliotheque Choisie*, 6 (1705), 376, translated in Jean Le Clerc, *The Life and Character of Mr John Locke*, trans. T.F.P. Gent. (London, 1706), 14; Maurice Cranston, *John Locke: A Biography* (London: Macmillan, 1957), 117, 361; Roger Woolhouse, *Locke. A Biography* (Cambridge: Cambridge University Press, 2007), 249.
33 Cranston, *Locke*, 312; Woolhouse, *Locke*, 266.
34 Pierre Des Maizeaux, ed., *A Collection of several Pieces of Mr. John Locke* (London: J. Bettenham, 1720), sigs. Aa5^{v}-6^{r}.
35 Des Maizeaux, *A Collection of Several Pieces*, 358–62 for 'Rules of a Society, Which met once a Week...' These were also published in *The Works of John Locke*, 10 vols. A new Edition, Corrected (London: T. Tegg et al., 1823, repr. Scientia Verlag Aalen: Germany, 1963), 10: 312–14.
36 MS Locke c.25 fols 56–57, Bodleian Library. This is in William Popple's hand (see note 39) and endorsed by Locke as 'Dry Club 92' (fol.57^{v}).
37 For a transcription of the 'Dry Club' manuscript (which I have used here) and comparison with Des Maizeaux's version of 'Rules of a Society', see John Locke, *Literary and Historical Writings*, ed. J.R. Milton in collaboration with Brandon Chua, Geoff Kemp, David McInnis, John Spurr, and Richard Yeo (Oxford: Clarendon Press, forthcoming).
38 MS Locke c.25, fol. 57^{r}, Bodleian Library, for the 'Dry Club'; see also 'Rules of a Society,' in Des Maizeaux, *A Collection of Several Pieces*, 359, for the question: '*Whether he loves all Men, of what Profession or Religion soever?*'
39 John Locke, *A Letter Concerning Toleration*, trans. William Popple (London: A Churchill, 1689). Cranston, *Locke*, 362 says that this stress on religious toleration was absent from 'the old Exeter House club and The Lantern.'
40 MS Locke c.25 fols. 56^{r-v}, Bodleian Library. See also 'Rules of a Society,' in Des Maizeaux, *A Collection of Several Pieces*, 360: 'That no Person or Opinion be unhandsomely reflected on; but every Member behave himself with all the temper, judgment, modesty, and discretion he is master of.'
41 MS Locke c.25, fol. 56^{v}, Bodleian Library.
42 Philip Milton, 'Pierre Des Maizeaux, *A Collection of Several Pieces of Mr. John Locke*, and the Formation of the Locke Canon,' *Eighteenth-Century Thought* 3 (2007): 283–84.
43 Benjamin Furly to Locke, 7/17 November 1692, *Correspondence*, 3:571. This point is noticed in Woolhouse, *Locke*, 274; and Milton, 'Pierre Des Maizaux,' 283.
44 Cranston, *Locke*, 117. Locke was elected FRS on 26 November 1668 as the 269th Fellow; see Michael Hunter, *The Royal Society and its Fellows 1660–1700: The Morphology of an Early Scientific Institution*, 2nd ed. (1982; Oxford: Alden Press, 1994), 184.

45 Damaris Masham to Jean Le Clerc, *Epistolario*, 2:492, 502, 508. Thomas Herbert (1656–1733), Eighth Earl of Pembroke, was the person to whom Locke dedicated his *Essay*.
46 On the variety of Locke's conversation partners, see Richard Yeo, 'John Locke on Conversation with Friends and Strangers,' *Parergon* 26, no. 2 (2009): 11–37.
47 John Locke, *Of the Conduct of the Understanding*, in *Works of John Locke*, 3:284.
48 Locke, *Conduct*, 287, 289. See Thomas Forster, ed., *Original Letters of Locke; Algernon Sydney; and Anthony Lord Shaftesbury* (London: J. B. Nichols and Sons, 1830), cxii for the claim that Locke himself suffered from 'nervous affections,' including 'false voices,' sounds, and images. On 'earworms,' see Michael C. Corballis, *The Wandering Mind* (Chicago: The University of Chicago Press, 2015), 4.
49 Locke, *Conduct*, 289.

Bibliography

Primary sources

Aristotle, *De Sensu et Sensibili*, 437a4–437a16. In *The Complete Works of Aristotle, the Revised Oxford Translation*, edited by Jonathan Barnes. 2 vols. Princeton, NJ: Princeton University Press, 1984.

Braithwaite, Richard. *Essaies Upon the Five Senses*. London: E. G[riffin], 1620.

Cervantes, Miguel de. *The Second Part of the History of Don Quixote of the Mancha*. Translated by Thomas Shelton. London: E. Blount, 1620.

De Beer, E.S., ed. *The Correspondence of John Locke*, 9 vols. Oxford: Clarendon Press, 1976–1989.

Des Maizeaux, Pierre, ed. *A Collection of Several Pieces of Mr. John Locke*. London: J. Bettenham, 1720.

Forster, Thomas, ed. *Original Letters of Locke; Algernon Sydney; and Anthony Lord Shaftesbury*. London: J. B. Nichols and Sons, 1830.

Heveningham, Colonel. 'A Song set by Mr. Henry Purcell, the Words by Colonel Heveningham.' *Gentleman's Journal, or, The Monthly Miscellany* 1 (1692): 7–29.

Hooke, Robert. 'A General Scheme, or Idea of the Present State of Natural Philosophy.' In *The Posthumous Works of Robert Hooke*, edited by Richard Waller, 1–70. London: S. Smith and B. Walford, 1705.

Le Clerc, Jean. *Epistolario*, edited by Maria Grazia and Mario Sina, 4 vols. Firenze: L. S. Olschki, 1987–1994.

Le Clerc, Jean. *The Life and Character of Mr. John Locke*. Translated by T.F.P. Gent. London: John Clark, 1706.

Locke, John. *An Essay Concerning Human Understanding*. 1690, edited by Peter H. Nidditch. Oxford: Clarendon Press, 1975.

Locke, John. *A Letter Concerning Toleration*. Translated by William Popple. London: A Churchill, 1689.

Locke, John. *Literary and Historical Writings*, edited by J.R. Milton in collaboration with Brandon Chua, Geoff Kemp, David McInnis, John Spurr, and Richard Yeo. Oxford: Clarendon Press, forthcoming.

Locke, John. *The Works of John Locke*. A New Edition, Corrected. 10 vols. London: T. Tegg et al., 1823, reprint, Aalen, Germany: Scientia Verlag, 1963.

Montaigne, Michel de. 'On the Art of Conversation.' In *The Complete Essays*. Translated by M.A. Screech. London: Penguin Books, 1991.

Secondary sources

Clark, Stuart. *Vanities of the Eye: Vision in Early Modern European Culture*. Oxford: Oxford University Press, 2007.

Corballis, Michael C. *The Wandering Mind*. Chicago: University of Chicago Press, 2015.

Cranston, Maurice. *John Locke: A Biography*. London: Macmillan, 1957.

Frangenberg, Thomas. 'Auditus Visu Prestantior: Comparisons of Hearing and Vision in Charles de Bovelles's Liber de Sensibus.' In *The Second Sense: Studies in Hearing and Musical Judgement from Antiquity to the Seventeenth Century*, edited by Charles Burnett, Michael Fend, and Penelope Gouk, 71–89. London: Warburg, 1991.

Harvey, Ruth E. *The Inward Wits: Psychological Theory in the Middle Ages and the Renaissance*. London: Warburg Institute, 1975.

Held, Richard, Yuri Ostrovsky, Beatrice de Gelder, Tapan Gandhi, Suma Ganesh, Umang Mathur, and Pawan Sinha. 'The Newly Sighted Fail to Match Seen with Felt.' *Nature Neuroscience* 14, no. 5 (2011): 551–53.

Hunter, Michael. *The Royal Society and Its Fellows 1660–1700: The Morphology of an Early Scientific Institution*. 2nd ed. Oxford: Alden Press, 1994.

Kim, Jiye G., Andrew J. Goldman, and Irving Biederman. 'Blind or Deaf? A Matter of Aesthetics.' *Perception* 37, no. 6 (2008): 949–50.

Milton, Philip. 'Pierre Des Maizeaux, *A Collection of Several Pieces of Mr. John Locke*, and the Formation of the Locke Canon.' *Eighteenth-Century Thought* 3 (2007): 255–91.

Thomas, Keith. *The Ends of Life: Roads to Fulfilment in Early Modern England*. Oxford: Oxford University Press, 2009.

Woolhouse, Roger. *Locke. A Biography*. Cambridge: Cambridge University Press, 2007.

Yeo, Richard. 'John Locke and Polite Philosophy.' In *The Philosopher in Early Modern Europe: The Nature of a Contested Identity*, edited by Conal Condren, Stephen Gaukroger, and Ian Hunter, 254–275. Cambridge: Cambridge University Press, 2006.

Yeo, Richard. 'John Locke on Conversation with Friends and Strangers.' *Parergon* 26, no. 2 (2009): 11–37.

Yeo, Richard. *Notebooks, English Virtuosi, and Early Modern Science*. Chicago: University of Chicago Press, 2014.

2 Awkward silences

John Barrell

In the last 30 years or so, the social history of the eighteenth century in Britain has come to be dominated by a single story: how the provincial gentry, and how the commercial middle class, fashioned themselves into members of a polite society; how they became polished, in all the senses the word brought with it from Italy and Paris – not grubby but cleaned up, not rough round the edges but smooth, not varnished but with a deep down lasting shine. This development has given us a new and valuable understanding of eighteenth-century British culture, especially of literature and the visual arts, but for dinosaurs like me, brought up on E. P. Thompson and history from below, it has not been a wholly positive development. It has relegated the vast majority of the people who lived in the eighteenth century to mere extras, walk-on parts in the corners of drawing rooms or street scenes, a process facilitated and confirmed by the recent revival of the word 'Georgian' in art exhibitions and in the version of British history peddled by the BBC's Channel 4.[1] Nobody would attempt to deny that there were poor people in the eighteenth century, but they had no claim on the elegant identity 'Georgian,' so we can hardly complain when they are cleansed from the historical record of the 'Georgian' period.

Of all the various sounds of civility, none was more civil, in Georgian Britain, than the sound of polite conversation. The wealth a man acquired might earn him the right to *aspire* to be regarded as a gentleman, but the surest way to fulfil that aspiration was by success in the art of polite conversation; indeed, the main point, as David Solkin has put it, of participating in that activity, 'was simply to demonstrate that one was qualified to do so.' [2] It is not surprising then that the genre of painting that has been taken to be especially characteristic of the Georgian age is the 'conversation piece.'[3] The phrase, as we shall see, is ambiguous and arguably misleading, but it will be helpful to begin with a short definition, and that is what Wikipedia is for. 'Conversation piece,' Wikipedia tells us,

> is a term for an informal group portrait, especially those painted in Britain in the 18th century, beginning in the 1720s. They are distinguished by their portrayal of the group apparently engaged in genteel conversation or some activity, very often outdoors. Typically the

> group will be members of a family, but friends may be included, and some groups are of friends, members of a society or hunt, or some other grouping. Often the paintings are relatively small, about the same size as a half-length portrait but in horizontal or 'landscape' format; others are much larger.[4]

In this essay, I shall be focussing first mainly on pictures of families and then on groups of gentlemen. What I say about family portraits will mainly be concentrated on the first half of the century when the conversation piece is more evidently an aspirational genre, often used to propose or to demonstrate that a family has attained a level of politeness which justifies the decision to commission such a picture. William Hogarth's portrait of Sir Andrew Fountaine and others (Figure 2.1) is a perfect case in point, the model to which painters and postulants to politeness both aspired. It is believed to show Fountaine, a famous art collector, regarded in his lifetime as a paragon of politeness, being invited by the auctioneer and art dealer Christopher Coke to add a new work to his collection. Coke points out its merits to Fountaine, and on the left, Mrs Coke, in a restrained brown silk dress, is also pointing out to Mrs Elizabeth Price something in the picture that has caught her eye.

Figure 2.1 William Hogarth, *Conversation Piece (Portrait of Sir Andrew Fountaine with Other Men and Women)*, c.1730–35, oil on canvas, 47.6 × 58.4 cm, Philadelphia Museum of Art: The John Howard McFadden Collection, 1928.

It is of the nature of the genre, aspirational as it often is, that the ideal of politeness is a bit more out of reach. Joseph Van Aken's *An English Family at Tea* (Figure 2.2) was painted around 1720, according to historians of dress and of domestic silver. Van Aken was one of the numerous Flemish artists who came to England to practise, drawn by the booming British economy and the lack of competent native artists. The rise of the notion of politeness in the historiography of the eighteenth century has made this painting quite well known. It is not known, however, whom it represents, though Solkin, the best guide to the painting of politeness, has pointed out that these men and women must have been real people: 'only a commissioned likeness would have required so highly individualised a representation of facial features and material possessions.'[5] The paraphernalia of the tea-table, for example, is painted with a precision which suggests that the family gathered round the table are anxious to show off their wealth and taste. Still, the fact that we do not know who they are suggests that they were not an old family, and that their new wealth did not last long. Perhaps they were casualties of the bursting of the South Sea Bubble.

There is much more to say about this picture, but no time to say it. I want instead to glance still more quickly at a few more family conversation pieces of the early Georgian period. William Verelst was a Dutch

Figure 2.2 Joseph Van Aken, *An English Family at Tea*, c.1720, oil on canvas, 99.4 × 116.2 cm, © Tate, London 2016.

painter who like Van Aken was attracted to Britain by the likelihood of employment offered by new money. In his conversation piece of 1741, of the Gough family at tea (Figure 2.3), the taste and wealth of the sitters is once again displayed – but not too pushily – by their sumptuous dresses, brand-new carpet, costly and tasteful tea set. Gough had made his money in the China and India trades and was now establishing himself as a country gentleman. In 1740, another Flemish painter, Joseph Francis Nollekens (Figure 2.4), painted a large group with, at its centre, the banker Sir Richard Child, well on his way to being raised to the peerage as the first Earl Tylney. He is accompanied by members of his family and by what are probably a few of his friends, even of his employees. William Hogarth began his career as a painter by producing 'small family pieces and conversations,' as George Vertue called them,[6] for example, his group portrait, of about 1730, of the fairly minor gentry family of Robert Jones (Figure 2.5), assembled for the occasion in the park surrounding the castle they lived in, visible on the horizon. Probably in the early 1730s, Joseph Highmore painted a conversation thought to portray the artist's family (Figure 2.6). Mrs Highmore, if it is she, sits at the centre of the picture, her work basket conveniently and unpretentiously to hand; her daughter Susanna sits at her feet, playing with

Figure 2.3 William Verelst, *The Gough Family*, 1741, oil on canvas, 76.2 × 85.8 cm, © private collection, reproduced by kind permission of the owner/Victoria and Albert Museum, London.

Figure 2.4 Joseph Francis Nollekens, *A Family in a Palladian Interior ('The Tylney Group')*, 1740, oil on canvas, 86.3 × 109.2 cm, private collection/Bridgeman Images.

Figure 2.5 William Hogarth, *The Jones Family Conversation Piece*, 1730, oil on canvas, 72.0 × 91.8 cm, National Museum Wales/Bridgeman Images.

Figure 2.6 Joseph Highmore, *Conversation Piece, Probably of the Artist's Family*, c.1732–35, oil on canvas, 64.7 × 77.5 cm, private collection/ Bridgeman Images.

a doll and drawing; and her son Anthony wears the negligent undress affected by poets and painters in portraits of them – Highmore himself, standing at the door, had hopes that both his children would become artists. The portrait by Francis Hayman of Jonathan Tyers (Figure 2.7), the entrepreneur and impresario of Vauxhall Gardens, sitting between his daughter Elizabeth and her fiancé John Wood, is assumed to be a betrothal portrait, and it embodies as well as any other picture in the genre the combination of informality and elegance which was the ideal of polite behaviour. Father and daughter are not too proud to sit on the ground, and her pug is not so well-trained as to have suppressed its natural friskiness. Wood, on the other hand, has taken up 'a pose,' as Elizabeth Einberg put it, 'of impeccable eighteenth-century elegance' but without the self-conscious stiffness that would have frozen the warmth of his smile.[7]

As everyone knows who has spent any time reading Boswell's account of how he spent his time in London, what was meant by conversation in the eighteenth century was not exactly what we mean by it now. Even when unqualified by the word 'criminal,' 'conversation' could mean 'having sex': when Boswell tells us he has had 'conversation' with the sex workers he finds in the streets, he does not mean to suggest they spent much time chatting. More generally, 'conversation' could refer to all

Figure 2.7 Francis Hayman, *Jonathan Tyers, with His Daughter Elizabeth, and Her Husband John Wood*, 1750–52, oil on canvas, 99.1 × 86.4 cm, Yale Center for British Art, Paul Mellon Collection.

kinds of social as well as sexual intercourse; one's 'conversation' could mean one's habitual circle of friends. But whatever the connotations of the word, at least when it does not mean having sex, it certainly involved a lot of talk. You could not converse with people in the mid-eighteenth century without *having conversation* with them; you could not have conversation without *conversing*. Whether or not conversation pieces – or, to give them their more usual eighteenth-century name, 'conversations' – were understood primarily as paintings of people engaged in spoken conversation, they were paintings of people who could not be imagined as *having conversation* without actually *conversing*. When Wikipedia tells us that conversation pieces 'are distinguished by their portrayal of the group apparently engaged in genteel conversation *or some activity*,' it does not, I take it, mean that the group can either be conversing or keeping quiet. I can think of odd 'conversations' in which the group may be involved in silently reading the bible, others where a silent family is listening to a reading by their paterfamilias or by a clergyman. But the only activity frequently depicted in conversation pieces that does not necessarily involve continuous conversation is making and listening to music: 'musical conversations,' as they were called.

Neither the Oxford dictionary nor historians of art seem to have recognised the fact, but the phrase 'conversation piece' was far more commonly used in the eighteenth century to describe a literary genre than an artistic one – twice as often, my electronic researches suggest. 'Conversation pieces' as they appeared under that name in periodicals, or as chapters in novels, were literary dialogues which were sometimes distinguished, however, as being less formal than dialogue itself, the genre or mode favoured of philosophers, from Plato to Bishop Berkeley. 'Some subjects are of so abstruse a nature as to be unfit for conversation pieces,' wrote one critic; according to another, the conversation piece was far more readable than 'the dullness of dialogue, the frigid elegance of narrative, or the formality of logical disquisition.'[8] I find it intriguing therefore that although the phrase associated painted conversation pieces so closely with spoken dialogue, and however informal they were taken to be, they very rarely went so far as to show their subjects actually talking to each other, when to show them engaged in polite conversation with each other seems so essential a part of their purpose. In contrast with the quiet exchanges and relaxed gestures in the Fountaine conversation piece, mouths seem wired shut. Sometimes, like Jonathan Tyers, one person looks towards another as if about to open a conversation, but this glance meets no reciprocal look, no encouragement. Very often the members of the group simply sit round and stare silently at the camera, each posing for his or her individual portrait, like the Thistlethwayte family by Thomas Hudson (Figure 2.8), or the gentlemen in this group portrait by Hayman – though here, evidently, there is no intention of representing the group in conversation (Figure 2.9), and the same is true of the quiet family in an anonymous group portrait at the Courtauld Institute, patiently waiting to be told they can relax (Figure 2.10). Sometimes, as in the Van Aken painting, one person or another stares out of the picture apparently at nothing in particular, as if actively to discourage a conversational opening from the others. Sometimes the gesture of a hand will suggest that someone is speaking, but the expressions of the presumed speaker and auditors will be without any of the animation we associate with conversation, as in William Verelst's Gough family, or Gawen Hamilton's picture of the Earl of Strafford and his family (Figure 2.11). The central gentleman in the alfresco sing-song by Nollekens (Figure 2.12) seems to have failed to persuade anyone to look either at him or at whatever he is pointing to; he would have attracted more attention, perhaps, had he remembered

Figure 2.8 Thomas Hudson, *The Thistlethwayte Family*, c.1758, oil on canvas, 184.7 × 214.6 cm, Yale Center for British Art, Paul Mellon Fund.

Figure 2.9 Francis Hayman, *Portrait of a Group of Gentlemen, with the Artist*, c.1740–45, oil on canvas, 111.8 × 142.2 cm, Yale Center for British Art, Paul Mellon Collection.

Figure 2.10 Arthur Devis, *Family Group*, 1740–1745, oil on canvas, 89.9 × 120.7 cm, The Samuel Courtauld Trust, The Courtauld Gallery, London.

Figure 2.11 Gawen Hamilton, *Thomas Wentworth, Earl of Strafford, and His Family*, 1732, oil on canvas, 93.4 × 83.3 cm, National Gallery of Canada, Ottawa.

Figure 2.12 Joseph Francis Nollekens, *Portrait of a Family,* 1740, oil on canvas, 127 × 101.6 cm, Yale Center for British Art, Paul Mellon Collection.

to open his mouth and speak. Compare the animated hubbub in some of Zoffany's large groups, painted in the 1770s (Figure 2.13), where busy argument seems as much at stake as politeness, in cheerful violation of the warnings given by de Piles, Richardson, Hogarth, Winckelmann and Algarotti against narrative ambiguity in pictures, like rooms where all are speaking at once;[9] or compare the polite but animated exchange of words and glances represented by John Sanders in the Pump Room at Bath in the early 1780s (Figure 2.14).

Children will sometimes (Figure 2.15) break the silence of family conversation pieces, sometimes by boisterous horseplay, but as often as not they, like the adults, are dumb, about to speak, perhaps, but not speaking, like the artist Joseph Highmore's own children, or like the

Figure 2.13 Johan Zoffany, *The Tribuna of the Uffizi*, 1772–77, oil on canvas, 123.5 × 155.0 cm, Royal Collection Trust/© Her Majesty Queen Elizabeth II 2016.

Figure 2.14 John Sanders, *The Pump Room, Bath*, 1780–85, oil on canvas, 76.0 × 96.0 cm, © Victoria Art Gallery, Bath and North East Somerset Council/Bridgeman Images.

Figure 2.15 William Hogarth, *The Cholmondeley Family*, 1732, oil on canvas, 71.0 × 90.8 cm, private collection/Bridgeman Images.

strange children of stranger parents painted by Arthur Devis, in *A Family Group on a Terrace in a Garden*, which I take to be a prophetic image of the imminent takeover of Georgian Britain by zombies (Figure 2.16). Towards the end of the century in particular, when these family groups begin to be invested in the representation of sentimental affection as well as of politeness, infants especially are allowed to break the silence. In the American artist John Singleton Copley's portrait of his family (Figure 2.17), the youngest children are no doubt to be imagined as clamouring for attention but seem to attract only the silent interest of the grown-ups, while the eldest child has already been socialised into the polite regime of silence. Or in the image of the family of Lord and Lady Willoughby de Broke, again by Zoffany, the great exponent of the group portrait, one of the children threatens to elicit a rebuke from his father for helping himself (I think it is a boy) to an extra biscuit, but whether the raised finger of his father is admonishing him for being greedy or noisy is unclear (Figure 2.18).

There are some obvious considerations that may have contributed to the strange effect of silence created by so many conversation pieces, of families in particular but also as we shall see of men in social – I hesitate to say sociable – groups which must be imagined to have gathered for the purpose of spoken conversation. Such pictures are made by juxtaposing

Figure 2.16 Arthur Devis, *A Family Group on a Terrace in a Garden*, 1749, oil on canvas, 101.5 × 124.5 cm, private collection/Bridgeman Images.

Figure 2.17 John Singleton Copley, *The Copley Family*, 1776–77, oil on canvas, 184.1 × 229.2 cm, National Gallery of Art, Washington D.C.

Figure 2.18 Johan Zoffany, *John, Fourteenth Lord Willoughby de Broke, and His Family*, c.1766, oil on canvas, 101.9 × 127.3 cm, J. Paul Getty Museum, Los Angeles.

a number of portraits for which the sitters sat individually, alone, not as a group, though it would not have taken a great deal of ingenuity to make their expressions and gestures more interactive. Portrait painters frequently engage their sitters in conversation to keep their expressions from freezing. To represent people speaking, to represent them so that we recognise them as doing so, requires their mouths to be open to some degree, but the French critic Du Fresnoy warns that 'the moving of the *Lips* ought to be but moderate':[10] anything more risks giving a grotesque permanence to transient expression, and to turn even those who rarely open their mouths to speak into people who never close them. This is the way to turn a polite portrait into vulgar caricature, as is apparent in a thousand graphic satires based on conversation pieces. *A Decent Story*, by James Gillray, makes the point perfectly (Figure 2.19). A cheerfully garrulous country squire is entertaining an army officer, a clergyman, and two women, all of them slightly the worse for drink, all of them open-mouthed, not quite vulgar, perhaps, but not quite as polite as they would like to be thought. Of the two country gentlemen in *A Toast for Old England*, by Thomas Rowlandson, after George Woodward (Figure 2.20), the one on the left, a dyed-in-the-wool Tory, a patriotic three-bottle man, no doubt, has opened his slack mouth to pronounce a drunken toast and will never close it again. No wonder people in so many polite conversations prefer to drink tea, the cup that cheers but not inebriates.

Figure 2.19 [James Gillray], *A Decent Story*, 1795, hand-coloured etching, 19.8 × 26.5 cm, private collection, photograph by Harriet Guest.

Figure 2.20 Thomas Rowlandson (after George Woodward), *A Toast for Old England*, 1798, stipple, etching and aquatint, 26.7 × 32.9 cm, © The Trustees of the British Museum.

Perhaps the main reason why spoken conversation seems to have been so rarely depicted in polite conversation pieces relates to the fact that the ability to engage in polite conversation, familiar as opposed to formal, and free as opposed to constrained, became so important as a test of gentility, too important to be left to chance, and apparently a great cause of social anxiety. By the heyday of the conversation piece, conversation had come to be treated as an art, and one that could be taught by rules. Rules for conversation were set out, often at great length, in conduct books and in periodicals throughout the century. Longish lists of such rules appear in the *Spectator* no. 557, the *Guardian* no. 24, the *Rambler* no. 173, the *Connoisseur* no. 138 and many more; and in practical manuals, often translated from French, with titles such as *The Conversation of Gentlemen*, *The Rules of Civility*, *The Art of Conversation*, and so on.[11] Lord Chesterfield tabulated and discussed some 30 rules for conversation, which were repeatedly reprinted; Charles Allen, claiming to write as a mother to her daughter, set out some similar rules in a manual called *The Polite Lady*; Erasmus Jones, keen to help the socially deprived, put together a longer account in *The Man of Manners, or, Plebeian polish'd*.[12]

Here, for example, is a digest of a short list of the rules published in the *Britannic Magazine*. Don't interrupt other people, but don't be put out when you are interrupted; let everyone speak in turn, and give everyone an opportunity to speak about whatever topic they know best; listen with patience, and answer with precision; don't trouble the company with your private concerns; avoid anecdotes; talk often, but never at length; don't start a second topic until the first is exhausted; don't laugh at your own jokes, and when the conversation is on a serious topic, don't interrupt with an ill-timed jest; don't whisper to your neighbour, or make remarks *sotto voce*, while the main conversation is continuing; don't say things about people in their absence that you would not say in their presence; don't speak of a man's virtues to his face or of his faults behind his back.[13] A man whose contribution to a conversation had a properly free and familiar air, achieved by obeying these rules, could apparently win the reputation of being a thorough gentleman, though I could easily have multiplied these rules many times over by summarising other such lists that appear in the writings of Lord Chesterfield and other manuals of conduct.

Some tables of regulations, however, went very much further, setting out the various rules to be observed when conversing with men of widely different characters. My favourite of these is by the German nobleman Baron Adolf Knigge, which, translated in 1799 by the minister of the Reformed German Congregation at the Savoy in London, covered 650 pages in two volumes, on *The Practical Philosophy of Human Life, or the Art of Conversing with Men*.[14] To some extent, this title is a good example of the capaciousness of the word 'conversation' in the eighteenth century; but most of the Baron's rules are about how to converse, in the modern sense of the word, with men of different characters.

Thus, Chapter 3 sets out, in 32 numbered paragraphs, rules for conversation with almost everyone, or rather every man, that it would be possible to meet in an average lifetime: with boasters, braggers, calumniators, deists, devotees, drunkards, enthusiasts, flatterers, freethinkers, lunatics, madmen, parasites, puffers, puritans, scoffers at religion, spendthrifts, villains, voluptuaries and votaries of other vices, and with people who are ambitious, arrogant, avaricious, cheerful, close, cunning, curious, 'eccentrical,' envious, forgetful, given to artifice, given to intrigue, good-natured, heedless, idle, imperious, imprudent, impudent, insidious, irascible, irritable, jealous, lazy, lively, melancholy, mistrustful, morose, obstinate, petulant, phlegmatic, revengeful, romantic, satirical, stupid, superstitious, suspicious, talkative, too modest, too timid, ungrateful, vain and weak. Chapters 4 and 5 set out the rules for conversation with people older than us, younger than us and the same age as us; with the very young and the very old, with our parents, our children, our relations in general; there are even 'a few words on old uncles and aunts.' Chapter 6 – and there are several more chapters after Chapter 6 – lays down the rules for a husband conversing with his wife throughout the duration of their married life, even when – if it should happen – they begin to find each other's company tedious, even when she finds herself attracted to another man and even when she is actually unfaithful.

The Baron intended this manual for the use of men, but not because he believed that the rules for the conversation of women are simpler, more self-evident, less in need of codification. On the contrary, 'ladies,' he writes,

> are restrained by many considerations that do not concern our sex. They depend more than men upon the opinion of the world, and must be more cautious and reserved in their conduct. On one hand they are indulged with more inadvertencies than our sex, and on the other hand with more whims: their conduct begins sooner to influence their character, while boys and youths may be more heedless without injuring themselves in a material degree; their experience is (or at least ought to be) confined chiefly to their domestic circle, whereas the man is tied more firmly to the state by his situation. For this reason many vices and virtues, acts and omissions, produce entirely different consequences if they be committed by one sex than if chargeable upon the other.[15]

A similar manual for women, the Baron concludes, would have to be more precise, more delicate and written by a woman. In fact, rules of conversation addressed to women are not very different from those addressed to men, because in both cases the rules are usually addressed to young people and are particularly concerned with warnings not to interrupt or contradict their elders. It is probably true, however, that rules against speaking scandal, and against seeking to appear 'smart and witty,'[16] are given greater emphasis where young ladies are being instructed.

But the rules most pertinent to the representation of conversation in pictures concern the gestures and expressions that speakers affect in the erroneous belief that by doing so they make their conversation more expressive and more pleasurable to their interlocutors. These include, according to the periodical *The Connoisseur*, those who converse 'chiefly with their arms and legs'; the 'Smirkers and Smilers,' who 'so prettily set off their faces, together with their words, by a *je ne-scai-quoi* between a grin and a dimple'; and the 'Attitudinarians' and 'Face-makers,' whose 'rules of eloquence are taken from the posture-master,' and who 'accompany every word with a peculiar grimace or gesture,' 'assent with a shrug, and contradict with a twisting of the neck; are angry by a wry mouth, and pleased in a caper of a minuet step.'[17] Once again, I will pass over numerous other examples; I have barely had space to list a small fraction of the endless number of the rules that govern the art of conversation.

Conversation, one conduct book announces, will give a 'certain ease and elegance of address, which is not to be obtained in any other way.'[18] And with so many lists of rules, so much instruction available, how could the aspirant after gentility possibly fail to become proficient as a polite conversationalist? But then again, is not 'ease of address' likely to have been the very last thing we could acquire by trying to learn, and to internalise, the vast body of the rules of conversation? As early as 1702, Simon Parker pointed out the problem posed by this multiplicity, which he understood as chiefly a result of the division of trades and professions, to each of which, he supposes, we need to address ourselves differently. 'What's to be done then?' he asks. 'How shall we contrive that our Rules of Conversation be neither too compendious nor too comprehensive?'

> How is it Possible to accommodate Rules and Maxims to such an infinite variety of Pretensions; If you espouse the Interest of This or That or T'other in opposition to the Rest, you mistake your Measures and your Argument: if you're for *Indulgence* and *Toleration*, you confound your whole Design, and bring the world back to *Babel-Colloquies*.[19]

Parker's answer is long and not very persuasive. The main point he has to make about the rules is that there are simply too many of them to remember. And they are almost all, in function if not in form, prohibitions, almost never permissions; they tell us what we may not do or say, not what we may. Anyone who took these rules seriously – and the sheer number of lists of rules and of treatises on polite behaviour suggests that many must have – would surely have been frozen, paralysed by the fear of making mistakes, of offending against the rules, of exhibiting themselves in public as anything but easy, informal, elegant, polite. Almost any expression or gesture represented in a conversation piece in order to represent spoken conversation would offend against one rule by following another. Almost any expression marked enough to indicate that a sitter was indeed speaking would be likely to seem to some judges excessive, exaggerated and lacking in breeding. Inevitably, sitters and artists felt more 'at ease,' less open to criticism of their manners, their taste, by reducing the signs of conversation to almost nothing – gestures that nobody seems to respond to, or even to notice – or to nothing at all.

This extraordinary image, by Charles Philips, shows a gathering of La Table Ronde, a club founded by Frederick, Prince of Wales (Figure 2.21), one of those where hierarchy was supposedly suspended, though Frederick is shown seated beneath the emblem of his status, the three feathers, which is carved above the portrait of Princess Anne wearing the same hunting uniform as the men wear.[20] In this club, Frederick apparently played Prince Hal as if in Henry IV Part 1, except that here he is formal where we might have expected him to be mischievous, bibulous and garrulous. Following his lead as they must, his jolly friends sit round without a hint of jollity, many of them more or less unaware of the presence of their fellow members. Anything less like the Boar's Head in Eastcheap would be hard to imagine.

In the course of some luminous pages about mid-century conversation pieces, Peter de Bolla focusses on sociability as conveyed by the looks and exchanges of looks among the figures in a painting, and between them and the spectators:

> Someone looks out of the image at us, and we look him or her back in the eye. Once in the image, as it were, the eye then follows the various sight lines of exchanged looks in a temporalizing and narrativizing movement. A good example of this is the painting by Hogarth [now attributed to Joseph Highmore] entitled *A Club of Gentlemen*, in which the figure closest to us on the left of the image looks at us [Figure 2.22]. We return his gaze and then follow the sequence of exchanged looks around the table. The activity of looking presented by the image gives the viewer the sensation of being included within the plane of representation, within the conversation. *A Club of Gentlemen* is extremely effective in conveying a kind of casual sense to this staging of the scene.[21]

To me, the 'sequence of exchanged looks' seems more or less unsequenced, which I take as appropriate to the image, for though the group

Figure 2.21 Charles Philips, *Frederick, Prince of Wales, with Members of 'La Table Ronde,'* 1732, oil on canvas, 128.0 × 102.2 cm, Royal Collection Trust/© Her Majesty Queen Elizabeth II 2016.

is of gentlemen, Highmore is not, I take it, overly concerned to represent their conversation as 'polite' – the chair kicked over by the man nearest us on the right, who seems the worse for drink, hardly suggests a group like Hayman's on its very best behaviour. At least two men may be speaking at once, but to whom? Who is listening? At whom are the gestures of the various men directed? This is a picture in which a number of rules of polite conversation are being casually broken.

Figure 2.22 Joseph Highmore, *A Club of Gentlemen*, c.1730, oil on canvas, 47.0 × 58.4 cm, Yale Center for British Art, Paul Mellon Collection.

One picture in particular seems to have come to be regarded by a number of art historians as the perfect image of polite conversation among men, in part no doubt because the social circle it represents includes Joseph Addison and Richard Steele, whose collaboration on the periodical the *Spectator* provided the model for male conversation-clubs. It dates from about 1711, is by James Thornhill, and currently has the title *Andrew Quicke in Conversation with the 1st Earl of Godolphin, Joseph Addison, Sir Richard Steele, and the Artist* (Figure 2.23). David Solkin and Jon Mee see this as an illustration of the kind of private conversation that created the Habermasian public sphere, in which the participants 'practise conversation as a form of polite exchange between equals that puts to one side hierarchy and inequality.'[22] They are not equals, of course: Godolphin was not only an Earl, but was probably, Solkin demonstrates, the patron of all these men, responsible for seeing that they were all possessed of comfortable sinecures. His pre-eminence is hinted at, Solkin suggests, by his nearly central position – he is the one wearing a red coat, gently swilling his wine in his glass, writing a note or memorandum, and, Solkin tells us, speaking. And to show that he is, actually, speaking, Solkin refers us to a close-up detail, which to my eye proves nothing, except that, if Godolphin is indeed speaking,

Figure 2.23 James Thornhill, *Andrew Quicke in Conversation with the 1st Earl of Godolphin, Joseph Addison, Sir Richard Steele, and the Artist*, c.1711–12, oil on canvas, 66.4 × 91.4 cm, private collection, photograph courtesy of the Paul Mellon Centre for Studies in British Art.

he has been caught at the moment of uttering a bilabial consonant, for his mouth is undoubtedly closed.[23] No one else appears to be speaking, except for what may be a murmured word passing between Addison, on the left, and the servant who has brought him a note; the transaction between them may be what has caught Godolphin's eye, and broken the flow of his speech; and if so, several rules of conversation have been broken at once: whispering when someone else is speaking, interrupting the speaker and seeming to notice that one has been interrupted. But all that would give the picture a drama it does not seem to deserve; for once again, this conversation, so-called, is between men who are paying each other very little attention. Like Godolphin, Steele, on the far right, may be looking at Addison, who is looking away from him; Quicke, leaning on the table, seems to be looking quite intently at Thornhill himself, the painter of this picture, who is looking intently at us. You could hear a pin drop.

Sometimes, however, the silence of conversation pieces seems to come across not as the mere absence of conversation but as vital to how they seem to invite us to understand them. I can try to make the point by comparing a pair of mezzotints after Richard Houston, of John Wilkes and members of the Society of Supporters of the Bill of Rights, meeting in London during the furore that followed the expulsion of John Wilkes from the House of Commons in 1769. One shows three city aldermen and members of parliament (Figure 2.24), James Townsend, standing, John Sawbridge, seated on the right and leaning on a copy of Locke on government, and William Beckford, who famously managed to combine a zeal for political liberty with the ownership of several slave plantations in the Caribbean. Sawbridge rests his chin in his hand and looks in an unfocussed way out of the picture, as though he has long since ceased to concentrate on what should be an important tactical discussion. Beckford looks out at us. In his hand is a petition detailing the peculations of Henry Fox, Lord Holland, which the trio have met to discuss.[24] Beckford's fingers seem to be quietly rapping the table, as if to recall the attention of the company to the business at hand; until he has their attention, he seems to have no desire to attend to any other proceeding. Townsend, meanwhile, has risen from the table and looks away from the circle as if he too has other things on his mind. This is a conversation piece, once again, about a conversation that is not happening; but like the portrait of Godolphin and his clients, it is not happening in a much more pointed way than in the family groups we looked at earlier. For these people have come together to talk, and to talk about issues vital to the constitution of Britain; but for whatever reason they are not talking.

The pendant however makes the characteristic silence of the conversation piece still more pointed (Figure 2.25). It shows John Wilkes himself in the centre, with, on the right, the radical parson John Horne Tooke,

Figure 2.24 Richard Houston, *Willm Beckford Esqr. ... Ja^s Townsend Esqr. ... John Sawbridge Esqr.*, 1769, mezzotint on paper, 30.4 × 39.0 cm, © The Trustees of the British Museum.

Figure 2.25 Richard Houston, *Mr Serjeant Glyn, John Wilkes Esqr, The Revd Mr John Horne*, 1769, mezzotint on paper, 30.5 × 39 cm, © The Trustees of the British Museum.

and on the left, the slightly less radical lawyer, John Glynn. Spread out between them on the table are more documents relating to the controversy over the publication of number 45 of the *North Briton* and the repeated expulsions of Wilkes from the House of Commons. These men are much more alert to each other's presence; they were, wrote Henry Angelo, in his account of what he recognises as 'a conversation piece,' 'sworn friends, and inseparable companions,'[25] and the sense of their mutual engagement is beyond anything we have looked at so far. The gesture of Tooke's left hand suggests he has just been speaking or is about to speak, but he and Wilkes are both turned towards Glynn, who seems to have gathered himself as if about to reply. He leans into the central space and in doing so seems already to have succeeded in winning their attention. As usual, nobody, it seems, is actually speaking, but this is a good deal nearer to a conversation, in our sense of the word, than most of the pictures we have looked at. It appears to be a picture, indeed, of a momentary pause in conversation, a moment of suspense, full of tension because full of silence.

Conversation pieces, and pictures described as 'conversations,' had once been very noisy, very boisterous even. In eighteenth-century Britain, the genre was regarded as having originated in the Low Countries, and the artists associated with the genre were mostly Dutch. My searches in *Eighteenth Century Collections Online* suggest that David Teniers was the artist most frequently referred to as a painter of conversations, with particular concentration on his images of drunken 'boors,' as they were known in Britain too, in pubs (Figure 2.26). Second most referred to was Hogarth, not particularly on account of his group portraits of middle-class families, but very largely because of his association with the noisy satire *A Midnight Modern Conversation* (Figure 2.27), rather more like the Boar's Head than the Prince of Wales's *Round Table.* Next came Adriaen Ostade, also a specialist in pubs and pub gardens, and Jean-Antoine Watteau, whose fêtes champetres were often referred to as conversations. Then came Egbert van Heemskirk, and then Adriaen Brouwer (Figure 2.28).

Towards the end of the eighteenth century, something like these boorish conversations began to be popular in Britain, but they were not images of the raucous, roaring peasantry of the kind who had appeared in Dutch conversations. Agricultural workers in Britain could in certain circumstances be seen, could be painted, but, as Peter Denney has shown, they were not to be heard, or only if their voices reached us from a safe distance.[26] Perhaps the chief painter of English boors was George Morland, whom I first wrote about 35 years ago in a book called *The Dark Side of the Landscape.* Of all Morland's paintings, the one I made the highest claims for was *Outside the Alehouse Door* (Figure 2.29), and I became fascinated by the language in which it had been described by contemporary critics, who seemed to respond to it as if to a threat.

Figure 2.26 David Teniers the Younger, *Tavern Scene*, 1658, oil on panel, 48.7 × 68.7 cm, National Gallery of Art, Washington D.C.

Figure 2.27 Unknown Artist (after William Hogarth), *A Midnight Modern Conversation*, c.1732, oil on canvas, 76.2 × 163.8 cm, Yale Center for British Art, Paul Mellon Collection.

Figure 2.28 Adriaen Brouwer (manner of), *Village Carouse*, c.1630, oil on panel, 39.5 × 52.5 cm, Rijksmuseum, Amsterdam.

Figure 2.29 George Morland, *Outside the Alehouse Door*, 1792, oil on canvas, 56.0 × 48.3 cm, © Tate, London 2016.

'Two peasants are regaling at the door of a cottage, and seem in earnest conversation. The figures are drawn with a force highly characteristic.' 'A group of English figures regaling themselves, which, like true sons of liberty, they seem determined on in spite of all opposition.'[27] I was intrigued that both critics used the word 'regaling,' which might be appropriate to Brouwer's Master Drinker, downing a whole half gallon on his own, but hardly applies to Morland's pair, with only a single pint of ale between them. The implication of 'regaling,' I take it, is that any alcohol at all is an extravagant treat where the poor are concerned, and that their money would be better spent on their families than on their own coarse pleasures. I was intrigued too by the nervous description of these men as 'true sons of liberty,' resisting the demand – again the implication is clear – that they should accept the status of mild, temperate, industrious, submissive labourers that the counter-revolutionary ideology of the 1790s had fashioned for them. And, of course, they should be quiet, and not 'in earnest conversation.' On the basis of the hints and anxieties implicit in these opinions, but also of the expressions and attitudes of the two men, I suggested that we were to understand them as talking radical politics, and that their conversation, if it could be heard, would have alarmed the great majority of those who saw the picture when it was made. But looking at it again, after writing this essay, it seems yet another image in which conversation is implied but not actually taking place. The sense of danger it communicates seems to derive not as I once imagined from what the two men are saying to each other, but from the ominous silence that seems to have fallen between them. Once again, a visible silence creates tension, either between the two men, or between them and us, as if they can resume their discussion only when we are out of earshot. It is less what the men are saying that is alarming than the fact that we cannot know what they are not saying.

One of the important arguments in Jon Mee's recent book, *Conversable Worlds*, is that by concentrating on the politeness of polite conversation in the eighteenth century, we have imagined that the chief function of conversation was to establish a mode of deportment that allowed men of differing opinions to defuse the tensions among them, to prevent arguments, converse peaceably and arrive at consensus. In that belief, he argues, we have allowed ourselves to overlook the fact that, at least in certain circles, polite conversation was as much about collision as about consensus. It was a place where differences of opinion could be aired and acknowledged, where disputants could agree to differ, or just plain disagree, in the interests of upholding what each firmly believed to be true. Conversation was the engine of progressive enlightenment as well as of the civilising process, an activity as likely to create tension as to smooth it away; and my discussion of conversation pieces may suggest that, in a genre that found it so difficult to represent spoken exchange, groups with apparently nothing to say to each other may be thought of as exemplifying conversation as consensus, while images showing active, tense silences may be thought of as representing conversation as collision.

To summarise: I have been suggesting that conversation pieces in eighteenth-century Britain have a problem with the representation of polite conversation, and I have offered two possible reasons for that. First, I have suggested that polite conversation, far from being a relaxed and informal activity, was for many people, especially for those not born to it, apparently a trial by ordeal, or associated, at the least, more with anxiety than with ease. As Simon Parker pointed out at the start of the century, the protocols that governed it may have been more likely to inhibit than to facilitate the activity, and this may have been the case also for artists trying to represent it. For them, many of them foreigners and strangers to British polite culture, many of them on the other side of the polite/vulgar divide, the fear of offending against the rules in representations of polite conversation may well have been equally inhibiting.

Second, I have suggested that the representation of polite conversation would have been subtly inhibited also by the familiar eighteenth-century anxiety about the mismatch between the representation of activity itself, by nature diachronic, and the synchronic nature of the visual arts. This issue, chewed over by so many late eighteenth-century theorists of the visual arts, from Lessing to Fuseli, was always discussed in relation to the rapidly transient expressions of extreme emotion – extreme pain, extreme grief – which, it was argued, come to look merely grotesque when frozen in a still image. But if this were true, the same would apply to the transient but not at all extreme expressions of those engaged in convivial conversation, as arguably becomes apparent in caricature. The strangeness of some eighteenth-century conversation pieces, I have tried to argue, comes from the juxtaposition of a number of individual portraits in a social situation in which they are unable to be shown as actively participating.

When I began to write this essay, I expected to find myself writing at some length about a well-known but still intriguing watercolour *Soulagement en Prison*, by the British radical caricaturist Richard Newton, made in 1793 when he was 17 (Figure 2.30), to me one of the most intriguing of eighteenth-century 'conversations.' It shows a large group of political prisoners and their visitors in Newgate in 1793. Iain McCalman, still the best commentator on the picture, describes it as an image of 'British Jacobin civility,' and of a radical counter-public sphere in the period of the French Revolution.[28] In the event, this essay has turned out not as

Figure 2.30 Richard Newton, *Soulagement en Prison*, 1793, watercolour, 38.5 × 59.0 cm, courtesy of the Lewis Walpole Library, Yale University.

a discussion of this picture and its relation to the eighteenth-century conversation piece, but at best as a prolegomenon to such a discussion. Where does it take us with Newton's watercolour? Almost nowhere, perhaps, in that it appears so unlike every other conversation piece we have glanced at, but perhaps also quite a long way. I have suggested that men engaged in conversation can easily be represented in active conversation in caricature, but only at the cost of politeness. In this image, however, a group of men, many of whom would not have been thought of at the time as belonging to the polite classes, seem to be drinking, and conversing, freely, without becoming grotesque or ceasing to appear polite. Glances are exchanged and held, mouths are opened, and those who do not appear to be involved in conversational exchanges seem to be listening with interest to those who are. There is no extravagance of gesture, but nor is gesture avoided for fear of its seeming extravagant. We seem to be witnessing conversation devoid of the various anxieties that I have suggested are associated with it in so many other images of conversation. To some degree, this is presumably a result of the class fraction to which these men belong, and the democratic politics to which they adhere. It is easy for them to appear polite, with all the calm, relaxed informality politeness supposedly involved, precisely because they have no particular anxiety to appear so. Equally important, however, is the register, social and artistic, of the image. It has no apparent artistic pretensions, no sense of participating in the quest for upward social mobility that characterises the history of the conversation piece. The image neither is nor is not a caricature. It has the informality of caricature without the grotesqueness or the vulgarity, and thus matches up perfectly to the ideal characteristics of polite conversation.

The more I have looked at this image, the more this middle register, this space between caricature, usually etched or engraved though this image is a watercolour, and its opposite, relatively formal portraiture, usually in oil or pastel, seems to need more investigation. Here is another example, from 1792, *Warm Thoughts about Matrimony on a Winter Evening* (Figure 2.31), a print discussed by Harriet Guest in her book *Unbounded Affections*.[29] It shows a woman apparently reading to two others from a book about marriage, or perhaps – so the motif of the cat and mouse suggests – taking the mickey out of the woman on the right, for her reluctance or eagerness to enter the married state. The evident informality of their relationship is made especially apparent by the fact that they have hitched up their skirts to allow the fire to warm not just their feet but their legs, and it allows the reading to be interrupted by the conversation it stimulates. For lack of any better term, I have started describing the register of such images as *Soulagement* and *Warm Thoughts* as the 'plain style.' It seems to develop in the decades around 1800, and its value is in the way it allows representations of social life largely without satire, and without formality and the anxiety attached to it.

Figure 2.31 Anon, *Warm Thoughts about Matrimony on a Winter Evening*, 1792, etching and stipple engraving on laid paper, 23.0 × 29.0 cm, courtesy of the Lewis Walpole Library, Yale University.

Notes

1 'Georgians Revealed: Life, Style, and the Making of Modern Britain,' at the British Library, London 2013–14; 'Georgians: Dress for Polite Society,' at the Fashion Museum, Bath, 2014; Lucy Worsley, 'The First Georgians,' BBC4, 2014.

2 David H. Solkin, *Painting for Money: The Visual Arts and the Public Sphere in Eighteenth-Century England* (New Haven and London: Yale University Press, 1993), 42.

3 There has been much excellent work on the topic of the conversation piece since the publication in 1971 of Mario Praz's *Conversation Pieces: A Survey of the Informal Group Portrait in Europe and America* (London: Methuen). In preparing this essay, I have relied particularly on Ellen G. D'Oench, *The Conversation Piece: Arthur Devis and His Contemporaries* (New Haven: Yale University Press, 1980); Elizabeth Einberg, *Manners and Morals: Hogarth and British Painting 1700–1760* (London: Tate Gallery, 1987); Marcia Pointon, *Hanging the Head: Portraiture and Social Formation in Eighteenth-Century England* (New Haven, CT and London: Yale University Press, 1993); Solkin, *Painting for Money* (for publication details see previous note); Peter de Bolla, *The Education of the Eye: Painting, Landscape, and Architecture in Eighteenth-Century Britain* (Stanford: Stanford University Press, 2003); Kate Retford, *The Art of Domestic Life: Family Portraiture in*

Eighteenth-Century England (New Haven, CT and London: Yale University Press, 2006) and Desmond Shawe-Taylor, *The Conversation-Piece: Scenes of Fashionable Life* (London: Royal Collection Enterprises, 2009). Many of the paintings glanced at in this essay are discussed at greater length in one or more of those works.

4 https://en.wikipedia.org/wiki/Conversation_piece, accessed 13 July 2015.

5 Solkin, *Painting for Money*, 66.

6 George Vertue, *Notebooks* (for 1729), 6 vols, *The Walpole Society* 22 (1934): 40.

7 Einberg, *Manners and Morals*, 162.

8 Anon, *An Essay on Dialogue; Particularly on the Application of That Form of Writing to Matters of Law* (London: no pub, 1777), 40; review of Rev. A. Brice, *Scriptural Facts; or, Annotations on the Divinity of the Messiah*, 1799, in *The New London Review; or, Monthly Report of Authors and Books* 14 (January 1799 to June 1800): 121.

9 See John Barrell: *The Political Theory of Painting from Reynolds to Hazlitt: 'The Body of the Public'* (New Haven, CT and London: Yale University Press, 1986), 32 and 345–46, n. 71.

10 *The Art of Painting: by C.A. du Fresnoy: with Remarks: Translated ... by Mr. Dryden*, 2nd ed. (London: Bernard Lintott, 1716), 161–62.

11 John Constable, *The Conversation of Gentlemen Considered in Most of the Ways, that Make their Mutual Company Agreeable, or Disagreeable. In Six Dialogues* (London: J. Hoyles, 1738); Antoine de Courtin, *The Rules of Civility; or, the Maxims of Genteel Behaviour* (London: R. Scott et al, 1703); Stefano Guazzo, *The Art of Conversation* (London: J. Brett, 1738). See also Samuel Richardson, *Letters Written to and for Particular Friends, on the Most Important Occasions* (London: J. Osborn et al, 1750), 16–18; Robert Heath, *The British Palladium; or, Annual Miscellany: for ... 1768* (London: T. Beckett and P.A. De Hondt, 1768), 21–22; Anon, *The School of Wisdom and Arts being a Complete Repository or What is Most Curious in Art and Nature* (Berwick: W. Phorson, 1783), 96–98.

12 Philip Dormer Stanhope, Earl of Chesterfield, *Lord Chesterfield's Advice to His Son, on Men and Manners: or, a New System of Education* (London: Richardson and Urquhart, 1775), 15–27, 148–55; Charles Allen, *The Polite Lady: or a Course of Female Education. In a Series of Letters, from a Mother to Her Daughter*, 2nd ed. (London: Newbery and Carnan, 1769); Erasmus Jones, *The Man of Manners: or, Plebeian Polish'd* (London: J. Roberts et al, 1737[?]).

13 *The Britannic Magazine* 3 (1795): 246–47.

14 Adolf Freiherr von Knigge, *Practical Philosophy of Social Life; or the Art of Conversing with Men: After the German of Baron Knigge*, trans. P. Will, 2 vols. (London: Cadell and Davies, 1794).

15 Von Knigge, *Practical Philosophy*, 1:73–74.

16 Allen, *Polite Lady*, 84, 89.

17 [George Colman], *The Connoisseur. By Mr. Town, Critic and Censor-General*, 2 vols. (London: R. Baldwin, 1755–6), 2: 829–34 (no. 138).

18 Allen, *Polite Lady*, 149.

19 Samuel Parker, *Essays on Divers Weighty and Curious Subjects* (London: N. Cox, 1702), 71.

20 Shawe-Taylor, *Conversation-Piece*, 87–88.

21 Peter de Bolla, *Education of the Eye*, 49–51.

22 Jon Mee, *Conversable Worlds: Literature, Contention, and Community 1762–1830* (Oxford: Oxford University Press, 2011), 48.

23 Solkin, *Painting for Money*, 43.

24 'London Petition Instead of punishing, conferring Honours on a Paymaster the Public Defaulter of unaccounted Millions.'
25 Henry Angelo, *Reminiscences of Henry Angelo, with Memoirs of His Late Father and Friends*, 2 vols. (London: Henry Colburn, 1828), 1:59.
26 Peter Denney, '"The Sounds of Population Fail": Changing Perceptions of Rural Poverty and Plebeian Noise in Eighteenth-Century Britain,' in *Experiences of Poverty in Late Medieval and Early Modern England and France*, ed. Anne M. Scott (Farnham: Ashgate, 2012), 295–311.
27 See John Barrell, *The Dark Side of the Landscape: the Rural Poor in English Painting 1730–1840* (Cambridge: Cambridge University Press, 1980), 111–12.
28 Iain McCalman, 'Newgate in Revolution: Radical Enthusiasm and Romantic Counterculture,' *Eighteenth-Century Life* 22, no. 1 (1988): 95–110.
29 Harriet Guest, *Unbounded Attachment: Sentiment and Politics in the Age of the French Revolution* (Oxford: Oxford University Press, 2013), 190–92.

Bibliography

Primary sources

Allen, Charles. *The Polite Lady: Or a Course of Female Education. In a Series of Letters, from a Mother to her Daughter.* 2nd ed. London: Newbery and Carnan, 1769.

Angelo, Henry. *Reminiscences of Henry Angelo, with Memoirs of His Late Father and Friends.* 2 vols. London: Henry Colburn, 1828.

Anon. *An Essay on Dialogue; Particularly on the Application of that Form of Writing to Matters of Law.* London: no pub, 1777.

Anon. *The School of Wisdom and Arts Being a Complete Repository or What is Most Curious in Art and Nature.* Berwick: W. Phorson, 1783.

Chesterfield, Philip Dormer Stanhope, *Earl of. Lord Chesterfield's Advice to his Son, on Men and Manners: or, a New System of Education.* London: Richardson and Urquhart, 1775.

Colman, George. *The Connoisseur. By Mr. Town, Critic and Censor-General.* 2 vols. London: R. Baldwin, 1755–6.

Constable, John. *The Conversation of Gentlemen Considered in Most of the Ways, That Make their Mutual Company Agreeable, or Disagreeable. In Six Dialogues.* London: J. Hoyles, 1738.

de Courtin, Antoine. *The Rules of Civility; or, the Maxims of Genteel Behaviour.* London: R. Scott et al., 1703.

Dufresnoy, Charles-Alphonse. *The Art of Painting: by C. A. du Fresnoy: With Remarks: Translated into English, with an Original Preface, Containing a Parallel between Painting and Poetry: By Mr. Dryden.* 2nd ed. London: Bernard Lintott, 1716.

Guazzo, Stefano. *The Art of Conversation.* London: J. Brett, 1738.

Heath, Robert. *The British Palladium; or, Annual Miscellany: for the Bissextile Year 1768.* London: T. Beckett and P.A. De Hondt, 1768.

Jones, Erasmus. *The Man of Manners: or, Plebeian Polish'd. Being Plain and Familiar Rules for a Modest and Genteel Behaviour, on Most of the Ordinary Occasions of Life.* London: J. Roberts 1737.

Parker, Samuel. *Essays on Divers Weighty and Curious Subjects.* London: N. Cox, 1702.

Richardson, Samuel. *Letters Written to and for Particular Friends, on the Most Important Occasions*. London: J. Osborn et al., 1750.
Vertue, George. Notebooks (for 1729). 6 vols. *The Walpole Society* 22 (1934).
von Knigge, Adolf Freiherr. *Practical Philosophy of Social Life; or the Art of Conversing with Men: After the German of Baron Knigge*, translated by P. Will, 2 vols. London: Cadell and Davies, 1794.

Secondary sources

Barrell, John. *The Dark Side of the Landscape: The Rural Poor in English Painting 1730–1840*. Cambridge: Cambridge University Press, 1980.
Barrell, John. *The Political Theory of Painting from Reynolds to Hazlitt: 'The Body of the Public*.' New Haven, CT and London: Yale University Press, 1986.
de Bolla, Peter. *The Education of the Eye: Painting, Landscape, and Architecture in Eighteenth-Century Britain*. Stanford: Stanford University Press, 2003.
Denney, Peter. '"The Sounds of Population Fail": Changing Perceptions of Rural Poverty and Plebeian Noise in Eighteenth-Century Britain.' In *Experiences of Poverty in Late Medieval and Early Modern England and France*, edited by Anne M. Scott, 295–311. Farnham: Ashgate, 2012.
D'Oench, Ellen G. *The Conversation Piece: Arthur Devis and His Contemporaries*. New Haven: Yale University Press, 1980.
Einberg, Elizabeth. *Manners and Morals: Hogarth and British Painting 1700–1760*. London: Tate Gallery, 1987.
Georgians: Dress for Polite Society. Bath: Exhibition at the Fashion Museum, 2014–16.
Georgians Revealed: Life, Style, and the Making of Modern Britain. London: Exhibition at the British Library, 2013–14.
Guest, Harriet. *Unbounded Attachment: Sentiment and Politics in the Age of the French Revolution*. Oxford: Oxford University Press, 2013.
McCalman, Iain. 'Newgate in Revolution: Radical Enthusiasm and Romantic Counterculture.' *Eighteenth-Century Life* 22, no. 1 (1988): 95–110.
Mee, Jon. *Conversable Worlds: Literature, Contention, and Community 1762–1830*. Oxford: Oxford University Press, 2011.
Pointon, Marcia. *Hanging the Head: Portraiture and Social Formation in Eighteenth-Century England*. New Haven, CT and London: Yale University Press, 1993.
Praz, Mario. *Conversation Pieces: A Survey of the Informal Group Portrait in Europe and America*. London: Methuen, 1971.
Retford, Kate. *The Art of Domestic Life: Family Portraiture in Eighteenth-Century England*. New Haven, CT and London: Yale University Press, 2006.
Shawe-Taylor, Desmond. *The Conversation-Piece: Scenes of Fashionable Life*. London: Royal Collection Enterprises, 2009.
Solkin, David H. *Painting for Money: The Visual Arts and the Public Sphere in Eighteenth-Century England*. New Haven, CT and London: Yale University Press, 1993.
Wikipedia. 'Conversation Piece.' Accessed 13 July 2015, https://en.wikipedia.org/wiki/Conversation_piece.
Worsley, Lucy. 'The First Georgians.' BBC4, 2014.

Part II

Sonic spaces of civility and incivility

3 'The bell, like a speedy messenger, runs from house to house, and ear to ear'

The auditory markers of gender, politics and identity in England, 1500–1700[1]

Dolly MacKinnon

Playing earwitness to the power of a bell's potential reach, Protestant Bishop Hugh Latimer in 1552 concluded that 'if all the bells in England should be rung together at a certain hour, I think there would be almost no place, but some bells might be heard there …'[2] So what then did these bells have to say? Impossible to contain, a peal of bells or a tolling bell was readily understood by those within earshot, for it was the auditory hallmark of a civil society. Bells offered sounds of hope and warning, life and death, religious conformity, diversity and dissent as well as national mourning, commemoration and celebration. Bells were a performative representation of the politics of everyday life making audible the mechanisms of gender and status divisions, as well as religious diversity and political continuity, change and for some a form of sonic static.

Cultural historians have begun to tune their ears to the sounds of bells, and turned their eyes to the voluminous archives that survive about them, their physical form, their dedicatory inscriptions, as well as their many and varied uses. This is because, as Alain Corbin has succinctly outlined in the context of nineteenth-century France, bell ringing

> offers us a privileged access to the world we have lost. If we can agree that landscape is a way of analysing space, of loading it with meanings and emotion, and of making it available for aesthetic appreciation, the landscape defined by various kinds of sound fits this definition particularly well.[3]

Notable works for early modern England include David Cressy's *Bonfires and Bells* (1989), which examined how torches and bells functioned as important forms of memory after the Reformation as the Protestant national calendar redefined the auditory realm within England and New England. Cressy argues that forms of commemoration and anniversary bell ringing changed in the wake of these religious reforms.[4] Yet ringing (whether abundant or constrained) was also the voice of every local parish that had a bell or bells to ring. David Postles's more recent work builds on those of Cressy and Corbin to focus specifically on 'the soundscape of social position,' and his analysis reflects the costs of bell ringing for the dead in select parishes in Leicestershire.[5] Yet any analysis of the gender of the sounds and their costs remains elusive. It is Vanessa Harding's brief but evocative description of bell ringing in London that articulates the communicative capabilities of bells, when she notes that 'the tone, the number of bells and the duration of ringing advertised the status of the person for whose benefit the bells were rung.'[6] Bells also resonated gender differences through the varied way a bell or ring of bells could be rung. According to Clare Gittings, 'the passing bell' in early modern England rang 'nine strokes for a man,' 'six for a woman' and 'three for a child.'[7] The bell ringing rules of the London Scholars, dating from 1702 described the specific way of ringing the 'Death Peal' that produced a prolonged haunting silence between the sound of each bell, when compared with normal ringing practices. After the burial, there was also a 'short Peal of round Ringing,' which was the ringing of the available bells in order from the highest pitch to the lowest pitch. What is more, the sound of the bell could also be altered or muffled by tying leather, an 'old hat' or thick fabric round 'the Ball of the clapper of each bell.' As a practice this form of muffled tolling was used at the funerals of bell ringers, as well as young men and young women.[8]

Adapting Jacques Attali's model of the political economy of noise for an early modern setting, I examine what sounds, noises and silences were generated for the dead, how bells resonated sounds of religious conformity, nonconformity, sonic iconoclasm and even sonic slander, and how gender distinctions were made audible through the different ways specific bells were rung. I incorporate examples of how the early modern world heard, misheard or ignored these sounds. The audiation of early modern ears picked up the subtleties of religious, political and gender differences that resonated in the selection of a bell or bells, and the way in which they were rung. Ultimately, this article is about reconstructing an 'archaeology of resonance' and also 'of marginalities,' where bell sounds can simultaneously represent order and chaos, where the sounds are welcome to the ears of the state and anathema to the ears of subversives and vice versa.[9]

While bell ringing was a ubiquitous practice, between 1500 and 1700, it was neither uniform nor exact, for a bell rung in one parish may ring slightly before or after a bell in the next parish, even when the sounds they generated were intended to mark a specific time, and common event. Bells marked the times of the day, and were cyclical and seasonal, adapting to the lengths of the days as they waxed in the summer and waned during winter. Each parish bell had a distinctive tone and marked the boundaries of a parish. Parish pride was vested in parish bells, for as a visitor to St Bartholomew's church, Tong, Shropshire in 1720 noted 'the inhabitants here boast of nothing more at present than a great bell famous in these parts for its bigness.'[10] This bell was recast from an earlier bell and a translation of the original Latin inscription read 'Henry Vernon – To The Glory of God Almighty, the Blessed Mary and St. Bartholomew[.] Henry Vernon Knight caused this bell to be made 1518.'[11] Even when the sound of a neighbouring parish bell spilled across the parish boundaries, it was indicative of a specific domain. The venerable Bede recounted the story of the death of the Abbess St Hilda (c.614–80), and

> one sister of a distant Monastery, as she was sleeping, thought she heard the well known Sound of that Bell, which called them to Prayers, when any of them had departed this Life ... no sooner [had she] heard this, than she raised all the Sisters and called them into the Church, where she exhorted them to pray fervently, and sing a *Requiem* for the Soul of their Mother.[12]

Parish bells sounded out the dominant religious and political ideological stance of individual parishes even when this was in conflict with the State. Therefore, the early modern soundscape was a patchwork of sonic compliance and deviance.

By the early sixteenth century, the function of bells in the church rituals of life and death was firmly established and clearly understood in England:

> Bells are ordained to call the ministers and people to the church in times convenient, admonishing them to come to the preaching of the word of God, the ministration of the sacraments, the divine service, and prayer in the church[es] for the time used, to give knowledge of our Christian brother or sister departed this world, that both we may call to our remembrance our own mortality and also be moved with charity to pray for them so departed.[13]

The selection of a bell or peal of bells, and the way in which they were rung for the dead audibly differentiated between royalty, for whom bells were rung for free, and commoners, for whom bells were rung for a fee. Royal births, deaths and processions around the country were common causes for parish bell ringing. Pealing bells also provided a sonic accompaniment that rippled over the landscape wherever the royal retinue traversed. The bells of Saffron Waldon, Essex, rang in 1444 for the visit of Queen Margaret, and this practice was well established by 1500.[14] English parish bells even rang for ritual events in the lives of European monarchy, and Catholic Popes. In 1539, Henry VIII commanded a funeral service be held for the wife of Emperor Charles V, and 'the bells were rung in all the parish churches of London from Saturday noon to Sunday at night.'[15] During the reign of Queen Mary I, the parish bells of St John the Baptist Bristol rang 'for our holy father the pope at a cost of

2 shillings.'[16] Failure to provide ringing for royalty received a fine with the parish paying for its silence.[17] The fate of the lives and deaths of royalty were of great significance concerning the religious and political stability of the nation, and bells rang for their triumphs and tragedies in urban and rural parishes across England.

Parishes knew the worth of a good bell. The distribution of bells across parishes and counties by specific makers is indicative of how a good bell rang as a sonic advertisement of the bell maker's craft into the ears of those who heard it. St Michael's Cornhill, London, had 'a peal of five bells,' and between 1429 and 1431, 'the tenor or a sixth bell' was added by William Rus, a 'goldsmith,' 'Alderman of London' and 'Sheriff.'[18] The bell was called 'Rus,' and it was the recasting of 'Rus' in 1588 that reveals the importance of the preservation of an individual bell's pitch in relation to the other bells within the ring that was a significant issue to be considered when any one bell was recast or replaced. In this instance, the parishioners were not happy with the 'greate bell ... wch have byn [used] insteade of Rus and not lyked.' On 'viewinge' the 'newe bell,' the churchwardens 'had wt us a musicion to sounde the bell.' They also:

> Paide [18 pence] to ye Waytes of the Cittie that took paynes to take the note of or belles & to goe ... to take the note of the newe bell then cast. [Finally they] Spent [a further 12 pence] upon a company (of) other musicions to take a ffurther noute of the same bell.[19]

When Rus was reinstalled, the bell rang true within the peal and to the liking of the parishioners' ears. The importance of tuning bells is evident in the Archdeaconry of Suffolk in 1637, which asked every parish about the number, state and condition of their bells and 'are they tuneable?'[20] Parishioners knew the sounds of individual bells, as well as peals of bells, for each indicated sounds of warning, death, celebration or folly. Different bells spoke in different ways across the landscape.

Between 1500 and 1700, what was considered treasonous ringing during one monarch's reign could be considered patriotic in another. During the Henrician Reformation of England, the church bells raised the rebellion against the King in Lincolnshire in 1536 in response to Henry's dissolution of the monasteries.[21] During Edward's reign, the Devonshire parishes rang to call those who still held allegiance to a Catholic England to rebellion in 1549 in defiance of Protestant church reforms. The Privy Council's response was swift and specific, muting and depleting parish bells as they 'ordered the removal of all but a single small bell from every church tower in Devon.'[22] In 1553/4, the bells of St Michael's Cornhill, London, rang in celebration of a Catholic victory and overthrow of Sir Thomas Wyatt the younger who had led some 3,000 men from Kent in protest against Mary I's marriage to Philip II of Spain and with the intention of placing Elizabeth, a Protestant, on the throne.[23] Even with the Elizabethan Protestant Church settlement, the nation still lacked, at a parish level, a uniform adherence to the state religion and reformed bell ringing practices. For example, in Yorkshire, public Catholic worship still persisted during the 1560s and 1570s, and at Tickhill, the parishioners shunned the Protestant parish church to attend mass held in a disused church.[24] In a parish near York, 'the minister regularly locked up the building when none' of his parish 'responded to the call of the bell.'[25]

Therefore, the effective use of bell sounds only worked as a form of communication because of its significance and purposeful contribution to or absence from the parish soundscape. Parishes automatically recognised the gender implications of royal peals because of the specific dates – red letter days – set aside in the annual calendar. Elizabeth I's accession day was 17 November, and each year churchwardens' accounts across the country would 'provide for ringing the Quines grace,' or 'paid to ringe upon the Queines day.'[26] Even after her death, parish bells rang in her memory. But while the bells in many parishes rang for Queen Elizabeth I during her life, more than one parish remained silent. At Colne Engaine in Essex, Richard Norton, the churchwarden was brought before the Church courts because 'A poor boy of the parish called John Spurgin of the age of 16 or 17 did shut him and others out of the church whereby they could not ring as they aut to.'[27] The ultimate responsibility for the ringing or silencing of bells was with the parish sexton and churchwardens.

The contemporary diary of Philip Julius, Duke of Stettin Pomerania, a visitor to London in 1602, gives an excellent insight into how bell ringing in the capital sounded to unfamiliar ears:

> On arriving in London we heard a great ringing of bells in almost all the churches, going on very late in the evening. We were informed that the young people do that for the sake of exercise and amusement, and sometimes they pay considerable sums as a wager, who will pull a bell the longest, and ring it in the most approved fashion. Parishes spend much money in harmoniously sounding bells, that one being preferred which has the best bells. The old Queen is said to have been pleased very much by this exercise, considering that it was a sign of the health of the people.[28]

So, the significance of a complete absence of bell ringing, on the rare occasions this happened, spoke volumes. The London soundscape of 1603 was silent as England waited to find out its religious and political fate. Father Weston heard:

> During those few days in which she [Elizabeth 1] lay dying beyond all hope of recovery, a strange silence descended on the whole city,

> as if it were under interdict and divine worship suspended. Not a bell rang out. Not a bugle sounded – though ordinarily they were often heard.[29]

When James VI and I was proclaimed King, Weston's audiation of the proclamation peals would have been at odds with those who heard a positive sound of Protestant succession and political certainty. On 24 March 1603, the bell ringers at Prescot, Lancashire, were paid for 'ringing upon our Kinges proclemacion.'[30] In 1605, the bells celebrated the failure of the Gunpowder Plot and the preservation of James VI and I and his parliament. St Margaret's Westminster rang the bells 'when the parliament house should have been blown up.'[31] Cratfield, Suffolk in 1653/4 paid 'unto the ringers upon the powder treason day' for ringing its five bells.[32]

The purposeful silences offered up by some parishes for the living King's coronation day were also problematic. Earls Colne, Essex, was resistant to the 'popish' ceremonial reforms of Archbishop William Laud, and the village made audible their disdain for the national church of James VI and I by having 'no ringing in their Church on the King's coronation day.'[33] Bell sounds could equally temporarily act as a form of sonic iconoclasm in stopping unwanted forms of religious practice. At Radwinter, Essex in 1642, the service 'was disrupted when a crowd of forty to fifty boys and girls rushed into the church and set the bells jangling,' suggesting that 'jangling bells had an obvious use in a parish where the minister's aggressive ceremonialism had led to accusations of popery.'[34]

Across England, the customary bell tolling for the dead occurred 'as soon as any hath departed this Life, the Bell belonging to the Parish he liv'd in, was immediately tolled, and for some Time.'[35] No longer tolling for the soul, the post Reformation Protestant passing bell asked those within earshot: ' ... art thou working in the Field, or grinding at the Mill? Remember then, when thou hearest the Sound of the Bell for one departing, that though put up they Prayers for him.'[36]

It is from an analysis of the ringing fees and durations of sounds for the six bells at St Michael's Cornhill, London in 1521, that we can see a clear indication of the sonic hierarchy and distinctions between individual bells and their social commercial value (Table 3.1). There was a designated passing bell called Mary that was more expensive to ring at night than during the day. The use of the bells singly and in combination indicated different sounds and different levels of social status in the soundscape of death. Whether the parish had one bell or many bells, the way those bells were used and the duration of the tolling resonated the status and gender of those for whom they rang.

In 1589, the St Michael's Cornhill fees for ringing the bells remained the same as they were in 1521. Mr Barbor paid 2s 8d for a knell using the Mary bell for his child. Mr Ffygges paid 8s for 'one afternoons knyll with Rus & the Peales for his Wyfe,' and Widdowe Wormaoll paid 6d

Table 3.1 St Michael's Cornhill, London: Costs Determined by Bell Use, Function and Time at Burials, 1521

Bell	*Six Bells in Use*	*One Hour Knell*	*Six Hour Knell*	*Day*	*Night*
6	Great bell called 'Russe,' with all the peals of all six bells 'to ye Dirige and Masse'		8s 8d		
5	'Myhill' [Michael] bell with all the peals of all six bells 'to the Dirige and Masse'		7s 6d		
4	Mary [Passing] Bell with all the peals of all six bells 'to the Dirige and Masse'		6s		
4	Mary Bell with a peal of all six bells 'to brynge the corse to Chyrch'	20d			
3	Trinity bell with a peal of three bells 'to brynge the corse to Chyrche'	12d			
2	bell 'next downeward' with a peal of three bells 'to brynge the corse to Churche'	8d			
1	leste bell with a peal of two bells	6d			
4	Passing [Mary] bell for every hour tolled			4d	8d

Source: Overall (1871).

for a knell using the 4th bell for her husband. Mr Levens denoting the relative status of servants in the household paid 6d for a knell with the 'small bell for his maide servaunte.'[37]

The small bell was also used for children. But a person from outside the parish was charged more for buying into the parish soundscape, as in 1595, when 16d was 'paid for the afternoones knell of a Duche gentlewoman.'[38] In London, the 1709 combined accounts for St Mary Woolnoth and St Mary Woolchurch Haw showed that if you did not come from the parish and wished to be buried there and accompanied by bells, the costs were doubled, as in the case of Michael Wilson, who paid £2 to be buried under the 'organ gallery,' 6s 8d moiety for the great bell, and 2s 8d for the cost of ringing.[39]

The churchwarden's accounts of 1609 for St Mary's Church, Chelmsford, Essex, indicate the costs of the passing bell, the knell, the grave and also the ringers.[40] For Chelmsford, the rates for ringing the four bells were as follows, with the different costs indicating different bells and duration of ringing:

Double knell 20d
Knell 12d
Knell 10d
Knell 8d
Knell 4d

The passing bell could be rung up to three times, at 4d per ring.

3 Passing bells 12d
2 passing bells 8d
1 passing bell 4d
'tolling out' 4d with one example of the duration being 1 hour.

By the early twentieth century at St Mary's Chelmsford, the gendered soundscape of the death knell comprised 3 × 3 for a man, 3 × 4 for a woman.[41]

Therefore, in 1609, the yeoman Vincent Sutton's burial costs included '3 passing bells' at 12d and a 'double knell' at 20d, with 12d for the sexton's part in ringing, and a coffin costing 2s. Although we do not know which of the Chelmsford bells were tolled, Agnes Kitteredy, alias Lattermore, Widow, had 'two passing bells' rung at 8d, 'her double knell' at 20d, with the additional cost of 12d for the sexton and 6d 'for vi ringers' for ringing the double knell; her coffin burial cost 5s 4d.[42] Marie, the 18-year-old daughter of Widow Lyver of Moulsham, was buried with 4d paid 'for towlinge,' 'for ringinge out' another 4d, a 'double knell' at 20d, 8d for her grave, and the ring may have been muffled. The Chelmsford bells also rang for members of the Ingram family during June 1609. The sexton was paid 4d 'for tolling one houre for Thomas Ingrams wife' who lay close to death in childbirth. Four days later, their unbaptised daughter was buried in a grave at the cost of 4d, and while there was a 4d knell rung, no passing bell was tolled for the unbaptised child. Those who died while excommunicated from the church were also not entitled to the use of the passing bell. Miraculously on 20 July, Ingram's wife, for whom the bell had tolled three weeks earlier, was well enough to be 'churched.'

Regarding the use of the passing bell in seventeenth-century England, Protestants did not have a uniform view on whether or not to toll for the dead, but neither did they have an exclusively Protestant soundscape. In early seventeenth-century Jacobean England, well-connected and powerful Catholics 'were accorded relative freedom in their religious practices,' provided they had taken the Oath of Allegiance (1606).[43] Therefore, in 1614, when Henry Howard, Lord Privy Seal and first Earl of Northampton, a prominent Catholic, died, he made elaborate provisions for the Catholic rituals and trappings of death and burial.[44] He paid for 'a bell to be tolled for one or two days prior to the funeral' in loud contravention of the 1563 'Elizabethan [Protestant] pronouncement against bell-ringing' that limited all but monarchy to 'single peals of only an hour in length.'[45] This noise in certain Protestants' ears was the sound of a faithful Catholic who in observing the rituals of death and burial had insisted on his funeral having all of the appropriate sonic accompaniments his parish bells could resonate. Church bells to many in the parishes still remained 'these old Chyming chimneys of the drunken whore of *Babylon*' and indicative of 'The Peoples Idols-Temples-Steeples-Bells' and the noise of an unreformed soundscape.[46]

For example, Protestant superstition at the use of the passing bell and its 'popish' overtones had prevented the passing bell from being rung at All Saints, Newcastle, between 1643 and 1655 at great financial cost to the parish. The parish so keenly felt the loss of revenue that on 21 January 1655, they set aside their theological opposition and reinstated not only the custom of ringing the passing bell but also the payment of an appropriate fee into the sexton's hands.[47] The bells as an auditory marker of the excess of death were unmistakable, though, and in July 1665, during the Great Plague of London, Samuel Pepys noted 'Sad news of deaths of so many in the parish of plague … The bell always going.'[48]

But how then did the intersection between gender, social status and piety at St Margaret's Church, Barking in Essex, in the early eighteenth century manifest itself? These sounds demonstrated by the specific bells' sounds offered up either for a fee or for free for the dead are indicative of the ways in which 'gender,' in the words of Joan Wallach Scott, 'is a primary way of signifying relationships of power.'[49] Equally a bell's silence could exclude some deaths – those considered unworthy or ungodly, such as those excommunicated, traitors or suicides – from the communal soundscape of a neighbourhood. Joyful ringing in celebration of a death considered in the national interest is also part of the power of bells when the bell ropes are firmly in the hands of the Church and State. Early modern England was a patriarchal society evident in the unequal power relations and interactions between women and men who either conformed with or consciously rejected their roles within a hierarchical godly society.[50] Taken from another context, Judith Butler's schema of the performative nature of seeing and experiencing gender is equally applicable and adaptable to the soundscape of the early modern world.[51] In order for gender to be performed as sound, there are three essential features necessary to allow these cultural sounds to operate: a public forum, such as a parish landscape; a suitable mechanism for the transmission of sound, such as a bell or bells being rung; and a group of participants to hear or ignore these gendered sounds within a parish landscape. Bells were the voices of the dead that spoke about gender, status and piety to the living. Local and national memory rang in the ears of parishioners. The early modern soundscape was a timeshare entity in which many but not all men, women and children gained access – albeit unequal – to the range of statuses and gender identities that individual bell pitches could resonate across the parish landscape. Focussing on the rare survival of the Sexton's accounts for St Margaret's Barking for the early part of the eighteenth century, this chapter offers an analysis of that soundscape for the dead.

According to Clare Gittings, the 'passing bell consisted of nine strokes for a man,' 'six for a woman and three for a child, followed by one stroke for each year of his or her age.'[52] As women were unequal in the eyes and ears of the Church, and the law, so they were in the soundscape of early modern death, for while the cost of ringing for women and men of the similar status was the same, the way the bells were rung differentiated the dead by the sounds of gender. A parish bell at Pucknowle, Dorset, dated 1629, proclaims the patriarchal reality of everyday life and death for 'He that vill purchas honours gayne mvst ancient lather [to make noise] still mayntayne.'[53] Gendered tolling was a persistent practice but one that was not uniform. In early twentieth-century Essex, the customary practice marking gender difference and age persisted in many parishes across the county, but the way bells were rung was not the same.[54] Gendered tolling most commonly comprised 3 × 3 [9] for a man, 3 × 2 [6] for a women and 3 × 1 [3] for a child, but a few parishes tolled four strokes for a man, four strokes for a woman and a single stroke for a child. Parishes also differed in their order and placement of the elements of gender and age in the toll sequence.

The 'Account of the fees received for knells and burials' kept by the sexton of St Margaret's Church, Barking, Essex in the early eighteenth century reveals the practice of status and gender hierarchy that operated for an entire peal of six church bells, numbered from six (the greatest bell) to one (the smallest bell).[55] By the early twentieth century, Barking's six bells were gone, having been recast first in 1746 and then replaced with eight bells. Therefore, the survival of the Sexton's accounts is invaluable, although the bells they related to are no longer extant. Between 23 September 1712 and 17 May 1715, a period of over two and a half years, on 216 occasions, one of the six bells was used to sound a death knell (Table 3.2).

The incomplete nature of the accounts does not make it possible to calculate costs for each year, but between 23 September 1712 and 28 August 1713, the sexton knelled the bell 53 times, with the greatest

Table 3.2 St Margaret's Church, Barking, Total Number of Knells, 23 September 1712–17 May 1715

Number of Bells	*Number of Times Bell was Rung*
6	39
5	15
4	46
3	44
2	47
1	25
Total	216

Source: Essex Record Office D/P81/1/32.

share of the soundscape belonging to the third bell, followed by bells 4, 6, 2, 1 and 5 (Table 3.3).

Of these 53 parishioners for whom the bell tolled, 20 were men, 32 were women and one was an unidentifiable person. The additional entries for poor women and men for whom the bells were not tolled have not been included. Over the course of an entire year, the parish death knell rang with the greatest concentrations in the autumn and the spring (Table 3.4). The least rung bell was bell 5, followed by the cheapest and smallest bells (1 and 2). The soundscape of the parish was dominated by those determined to sound their respective social status in the sonic hierarchy of the parish.

Individual bells indicated the sound selected to reflect both the status and gender of the person for whom it was tolled, by the way the bell was rung (Table 3.5). The bells were rung half muffled or fully muffled for royal deaths and the deaths of churchmen and of the parish bell ringers.[56]

Women and men could pay the same amount to have the same bell tolled, but the bells that tolled were rung in differentiated ways. The customary practice of ringing that marked gender differences and age meant that in death, as they had been in life, a death knell for a woman sounded different to the ears of the parish than that of a man because of what in the nineteenth century was called a 'teller.' The 'teller' was the

Table 3.3 Total Number of Knells per Bell, 23 September 1712–28 August 1713

Year	*Bell 1*	*Bell 2*	*Bell 3*	*Bell 4*	*Bell 5*	*Bell 6*	*Total*
Number of times rung	5	5	16	12	4	11	53
Percentage	9%	9%	30%	23%	8%	21%	100%

Source: Essex Record Office D/P81/1/32.

Table 3.4 Annual Distribution of Death Knells Rung in the Parish, September 1712–August 1713

Bell	*Sept*	*Oct*	*Nov*	*Dec*	*Jan*	*Feb*	*Mar*	*Apr*	*May*	*Jun*	*Jul*	*Aug*	*Total*
6		3	2		1		1	2		2			11
5			1				3						4
4		2	3				5	2					12
3	1	1	1	1	3	2	1		1	2	1	2	16
2						1		1	2	1			5
1		1	1		1		1		1				5
Total	1	7	8	1	5	3	11	5	4	5	1	2	53

Source: Essex Record Office D/P81/1/32.

Table 3.5 Annual Gender Distribution of Death Knells Rung on the Six Bells in the Parish, September 1712–August 1713

Bell	*Women*	*Men*	*Unknown*	*Total*
6	7	4	0	11
5	3	0	1	4
4	8	4	0	12
3	6	10	0	16
2	3	2	0	5
1	5	0	0	5
Total	32	20	1	53

Source: Essex Record Office D/P81/1/32.

component within the ring that specifically denoted the sex and/or age of the person for whom the bell knells.

For St Margaret's Barking between 4 October 1713 and 17 May 1715, we can calculate the maximum and minimum costs paid for each knell and burials for each bell:

Bell 6: £1 3s 4d < > 5s
Bell 5: 9s 2d < > 2s 6d
Bell 4: 2s < > 1s
Bell 3: 4d < > 2d
Bell 2: 2d
Bell 1: 2d

The greater the cost of the bell, the more significant the individual, the greater the bell, and the longer the duration of the tolling. One way to attempt to determine and confirm the status of an individual with the bells rung is to locate where they were buried in the church of churchyard. There is no surviving plan of burial locations for eighteenth-century Barking. However, the high cost of the sixth bell, and the relative infrequency with which it was rung, implies that this was the most significant bell to have tolled for a funeral. I was able to match the survival of two funeral monuments at St Margaret's Barking, with the individuals for whom the sixth bell was tolled. The first was Elizabeth Bennet, the widow of the Hon. Robert Bertie who died on 12 January 1712/13. The sixth bell was rung at her burial. Her monument was described as a 'marble Corinthian pilaster with drapery, urn and cartouche of arms' and it survives in the chancel of St Margaret's Barking.[57] Unfortunately no ringing costs are listed for Bennet's funeral. The other individual for whom the sixth bell was tolled was Gilbert Jones who died on 29 November 1713. His burial was accompanied by a knell from bell 6 at a cost of five shillings. A headstone for Jones survives in the churchyard.[58] The examples for St Margaret's Barking offer a different parish soundscape to that which David Postles found for St Martin's Leicester, where burials within the church were accompanied by the ringing of all the parish bells.[59] Unfortunately, none of these examples provide any details about the duration of the knell, nor if additional ringing accompanied the funeral and burial. The ringing of bells, however, was a lucrative venture for the sexton, who received the fees from the ringing of the passing bell. For Barking, the ringing and burial fees for October 1714 generated £1 10s 6d. Furthermore, the poor in Barking do not appear to have had bells rung for them. Thomas Brooks, however, had bell 1 tolled for free, suggesting a social entitlement recognised by the churchwardens (Table 3.6).

Purchasing a burial place within the village sacral space and soundscape were important indicators of the status hierarchy, gender divisions and religious ideology of the parish. The wealthy could purchase sounds and burial spaces that reflected their social standing, but the godly deserving poor were also given access to these ideological mechanisms

Table 3.6 Sexton's Accounts, Barking, Essex, Sample Month 2–30 October 1714

Day	*Person*	*M*	*F*	*Bell 1*	*Bell 2*	*Bell 3*	*Bell 4*	*Bell 5*	*Bell 6*	*Total*
2	Richard Aylett	1				2d				2d
3	Ann Winn		1			2d				2d
4	Ediff Hubbard		1						£1-3s-4d	£1-3s-4d
12	Mary Archer		1				1s			1s
13	Gersham Bulcher	1						2s 6d		2s 6d
18	John Stradling	1					1s			1s
22	Alice Taylor		1				1s			1s
23	Ann Stracey		1				1s			1s
24	Robert Chifford – Poor Chadwell Ward, therefore no bell and no fee	1								
29	Thomas Brooks	1			No fee					
	Elizabeth ward		1		2d					2d
30	Sarah Colins		1		2d					2d
Total	12	6	7	0	4d	4d	4s	2s 6d	£1-3s-4d	£1-10s-6d

Source: Essex Record Office D/P81/1/32.

at the cost of the churchwardens. In the records for St Mary's church, Warwick, in 1686, where the fifth bell was rung for free, for 'every person that dyeth, having not money to pay for the ringing of the 5th bell,' and if they had no one to ring it for them, then the clerk would ring.[60] The normal charges for ringing a passing bell for any death at St Mary's comprised two shillings and sixpence for the Great bell, one shilling for the seventh bell and sixpence for the fifth bell.[61]

Therefore, the performance of death knells and anniversary bell ringing for monarchs established and reaffirmed gendered and status differences within the parish soundscape. It was against this national wash of high-status sounds that some local parishioners of all sorts attempted to buy their share of the ears of their neighbours through the bells that were rung for them at their death. This chapter has briefly provided some ways of filling in the gap in the historiography about the function of sounds for burials and commemoration in the early modern world, and the gendering of sounds. The bell sounds of burials with their gender differentiation and status hierarchies reaffirmed in the assembled congregations the reality of gender differences that resonated the unequal power relations of the patriarchal household, signifying piety and excluding profanity through the power of bell sounds. Thomas Adams best described the passing bell's communicative capacity as 'a speedy messenger' running 'from house to house and ear to ear.'[62] What early modern men, women and children did hear in the bell or bells of their parishes were the competing sounds of death and celebration, religious conformity and religious defiance, and all through the purposeful status and gender demarcations of the sound of parish bells.

Notes

1 Thomas Adams, 'The Sinners Passing Bell,' in *The Workes of Thomas Adams Being the Summe of his Sermons, Meditations, and Other Divine and Morall Discourses. Collected and Published in One Intire Volume. With Additions of Some New, and Emendations of the Old. The Titles Whereof are Placed in the Beginning of the Booke: and a Table of the Principall Points, in the End* (London: Thomas Harper, 1630), 249.

2 George Elwes Corrie, ed. *Sermons by Hugh Latimer, Sometime Bishop of Worcester, Martyr, 1555* (Cambridge: Parker Society, 1844), 498.

3 Alain Corbin, *Village Bells: Sound and Meaning in the Nineteenth-Century French Countryside*, trans. M. Thom (New York: Columbia University Press, 1998), ix.

4 Cressy, *Bonfires and Bells*; John Craig, 'Psalms, Groans and Dog-Whippers: the Soundscape of Sacred Space in the English Parish Church, 1547–1642,' in *Sacred Space in Early Modern Europe*, ed. Will Coster and Andrew Spicer (Cambridge: Cambridge University Press, 2005), 104–23.

5 David Postles, '"Ringing out those bells": Death and the Social Order in Early-Modern Leicestershire,' *Transactions of the Leicestershire Archaeological and Historical Society* 80 (2006): 31–41.

6 Vanessa Harding, *Dead and the Living in Paris and London 1599–1670* (Cambridge: Cambridge University Press, 2002), 203.
7 Clare Gittings, *Death, Burial and the Individual in Early Modern England* (London: Croom Helm, 1984), 133.
8 Ernest Morris, *The History and Art of Change Ringing* (London: Chapman & Hall, 1931), 506–7.
9 Jacques Attali, *Noise: The Political Economy of Music*, trans. Brian Massumi (Manchester: Manchester University Press, 1985), 22.
10 *A Guide to the Church: The Collegiate Church of St. Bartholomew, Tong*, 6th Revised Edition (Lichfield: Matt Print Ltd., 1998), 19.
11 *A Guide to the Church*, 19.
12 Henry Bourne, *Antiquities Vulgares; or, the Antiquities of the Common People. Giving Account of Several of their Opinions and Ceremonies with Proper Reflections Upon Each of Them; Shewing Which May Be Retain'd and Which Ought To Be Laid Aside* (Newcastle, 1725), 2–3.
13 'A Book Concerning Ceremonies to be used in the Church of England,' cited in Cyril S. Cobb, *The Rationale of Ceremonial 1540–1543 with Notes and Appendices and an Essay on the Regulation of Ceremonial during the Reign of King Henry VIII* (London: Longmans, Green and Co., 1910), 30.
14 J. Charles Cox, *Churchwardens' Accounts from the Fourteenth Century to the Close of the Seventeenth Century* (London: Methven & Co., c1913), 214.
15 Harding, *Dead and the Living*, 262, and Cox, *Churchwardens' Accounts*, 218.
16 Harding, *Dead and the Living*, 262, and Cox, *Churchwardens' Accounts*, 218.
17 Cox, *Churchwardens' Accounts*, 215.
18 John Stow cited in *The Accounts of the Churchwardens of the Parish of St. Michael Cornhill, in the City of London from 1456 to 1608. With Miscellaneous Memoranda Contained in The Great Book of Accounts, and Extracts from the Proceedings of the Vestry, From 1563 to 1607*, ed. William Henry Overall (London: Alfred James Waterlow, 1871), Preface, ix and 225.
19 Overall ed. *The Accounts of … St. Michael Cornhill*, 179.
20 *Articles To Be Enquired of In the Ordinary Visitation of the Right Worshipfull Master Doctor Pearson, Archdeacon of Suffolke* (London: Felix Kingston, 1637).
21 M.E. James, 'Obedience and Dissent in Henrician England: The Lincolnshire Rebellion 1536,' *Past and Present* 48 (1970): 3–76.
22 Eamon Duffy, *The Voices of Morebath: Reformation and Rebellion in an English Village* (New Haven and London: Yale University Press, 2001), 145.
23 Cox, *Churchwardens' Accounts*, 218.
24 Christopher Haigh, 'The Continuity of Catholicism in the English Reformation,' *Past and Present* 93 (1981): 44.
25 Haigh, 'The Continuity of Catholicism,' 44.
26 F.A. Bailey, ed. *Churchwardens' Accounts of Prescot, Lancashire, 1523–1607* (Preston: Record Society for the Publication of Original Documents relating to Lancashire and Cheshire, 1953), vol. 104, 107–8, and 111.
27 Colne Engaine History Society, *Colne Engaine: The Story of an Essex Village* (Colchester: Colne Engaine History Society, 1992), 5.
28 Diary quoted in Morris, *The History and Art of Change Ringing*, 26.
29 William Weston, *William Weston: The Autobiography of an Elizabethan*, trans. Philip Caraman (London: Longmans, Green & Co., 1955), 222.
30 Bailey, ed. *Churchwardens' Accounts of Prescot, Lancashire*, vol. 104, 143.
31 Cox, *Churchwarden's Accounts*, 220.
32 By this date, St Mary's Cratfield had five bells made between 1585 and 1637. See *The Suffolk Ringers Guild*: www.suffolkbells.org.uk; L.A Botelho, ed.

Churchwardens' Accounts of Cratfield, 1640–1660, Suffolk Record Society, vol. 42 (1999), 96.

33 Essex Record Office (E.R.O.) D/ACA 48, *Archdeaconry of Colchester Act Book*, Earls Colne, 3 May 1632.

34 John Walter, '"Abolishing Superstition with Sedition"? The Politics of Popular Iconoclasm in England 1640–1642,' *Past and Present* 183 (2004), 79–124; 105–6.

35 Bourne, *Antiquities Vulgares*, 3.

36 Bourne, *Antiquities Vulgares*, 8.

37 Overall, ed. *The Accounts of … St. Michael Cornhill*, 178–9.

38 Overall, ed. *The Accounts of … St. Michael Cornhill*, 185.

39 Guildhall Library, London, MS 8107, 'The Booke of Accts of Mr Robt Low the Church Warden Relating to Ye United Parishes of St Mary Woolnoth & St Mary Woolchurch Haw, For Ye Year 1709.'

40 Cecil Deedes and H.B. Walters, *The Church Bells of Essex, Their Founders, Inscriptions, Traditions, and Uses* (Aberdeen: W. Jolly and Sons, 1909), 205. There were four bells in 1591.

41 Deeds and Walters, *The Church Bells of Essex*, 204–5.

42 E.R.O., D/P94/1/24 'The Accoumpt of all Marriages, Cristenings, and Buryalls,' 26 June 1609.

43 Jane Woodward, *The Theatre of Death: the Ritual Management of Royal Funerals in Renaissance England* (Woodbridge: The Boydell Press, 1997), 144.

44 Woodward, *The Theatre of Death*, 144.

45 Woodward, *The Theatre of Death*, 58 & 144.

46 S. Chidley, *To His Highness the Lord Protector, &c. and to the Parliament of England. Valiant Swordmen, and Honorable Gentlemen* (London: no pub, 1656). Unpaginated.

47 Bourne, *Antiquities Vulgares*, 6.

48 Gittings, *Death, Burial and the Individual*, 135.

49 Joan Wallach Scott, *Gender and the Politics of History* (New York: Columbia University Press, 1988), 44–45.

50 Dolly MacKinnon, 'The Godly Family of the Seventeenth Century and John Howard's Australia,' in *Future Imaginings: Sexualities and Genders in the New Millennium*, eds. Delys Bird, Wendy Were and Terri-Ann White (Crawley, WA: University of Western Australia, 2003), 101–16; 103–6. See also Dolly MacKinnon, *Earls Colne's Early Modern Landscapes* (Aldershot: Ashgate, 2014), 74–76, 249–52.

51 Judith Butler, *Gender Trouble: Feminism and the Subversion of Identity* (New York: Routledge, 1990), 139.

52 Gittings, *Death, Burial and the Individual*, 133.

53 Williams Andrews, ed. *Curious Church Customs and Cognate Subjects* (Hull: William Andrews & Co, 1895), 53–4.

54 Deedes and Walters, *The Church Bells of Essex*, *passim*.

55 E.R.O., D/P81/1/31, September 1712 to 17 May 1715.

56 Deeds and Walters, *The Church Bells of Essex*, 168.

57 *Royal Commission on Historical Monuments (England): Essex*, 4 vols. (London: His Majesty's Stationery Office, 1921), 2:6.

58 *RCHM (England): Essex*, 2:7.

59 Postles, '"Ringing Out Those Bells,"' 38.

60 Cox, *Churchwardens Accounts*, 212.

61 Cox, *Churchwardens Accounts*, 212.

62 Adams, *The Workes of Thomas Adams*, 224.

Bibliography

Primary sources

Adams, Thomas. *The Workes of Thomas Adams Being the Summe of his Sermons, Meditations, and other Divine and Morall Discourses.* London: Thomas Harper, 1630.

Andrews, William, ed. *Curious Church Customs and Cognate Subjects.* Hull: William Andrews & Co., 1895.

Archdeacon of Suffolk. *Articles to be Enquired of in the Ordinary Visitation of the Right Worshipfull Master Doctor Pearson, Archdeacon of Suffolke.* London: Felix Kingston, 1637.

Bailey, F.A., ed. *Churchwardens' Accounts of Prescot, Lancashire, 1523–1607.* Vol. 104. Preston: Record Society for the Publication of Original Documents relating to Lancashire and Cheshire, 1953.

Botelho, L.A., ed. *Churchwardens' Accounts of Cratfield, 1640–1660.* Vol. 42 vol. Woodbridge: The Boydell Press/Suffolk Record Society, 1999.

Bourne, Henry. *Antiquities Vulgares; or, the Antiquities of the Common People.* Newcastle: J. White, 1725.

Chidley, S. *To his Highness the Lord Protector, &c. and to the Parliament of England. Valiant Swordmen, and Honorable Gentlemen.* London: S.N., 1656.

Cobb, Cyril S. *The Rationale of Ceremonial 1540–1543 with Notes and Appendices and an Essay on the Regulation of Ceremonial during the Reign of King Henry VIII.* London: Longmans, Green & Co., 1910.

Corrie, George Elwes, ed. *Sermons by Hugh Latimer.* Cambridge: Parker Society, 1844.

Cox, J. Charles. *Churchwardens' Accounts from the Fourteenth Century to the Close of the Seventeenth Century.* London: Methven & Co, 1913.

McClure, N.E., ed. *The Letters of John Chamberlain.* 2 vols, Memoirs XII, Part I, The American Philosophical Society. Philadelphia: The Lancaster Press, 1939.

Overall, William Henry, ed. *The Accounts of the Churchwardens of the Parish of St. Michael Cornhill, in the City of London from 1456 to 1608. With Miscellaneous Memoranda Contained in The Great Book of Accounts, and Extracts from the Proceedings of the Vestry, From 1563 to 1607.* London: Alfred James Waterlow, 1871.

Secondary sources

Anon. *A Guide to the Church: The Collegiate Church of St. Bartholomew, Tong.* 6th ed. Lichfield: Matt Print, 1998.

Attali, J. *Noise: The Political Economy of Music.* Translated by Brian Massumi. Manchester: Manchester University Press, 1985.

Butler, Judith. *Gender Trouble: Feminism and the Subversion of Identity.* New York: Routledge, 1990.

Colne Engaine History Society. *Colne Engaine: The Story of an Essex Village.* Colchester: Colne Engaine History Society, 1992.

Corbin, Alain. *Village Bells: Sound and Meaning in the Nineteenth-Century French Countryside*. Translated by M. Thom. New York: Columbia University Press, 1998.

Cressy, David. *Birth, Marriage, and Death: Ritual, Religion, and the Life-Cycle in Tudor and Stuart England*. Oxford: Oxford University Press, 1997.

Cressy, David. *Bonfires and Bells: National Memory and the Protestant Calendar in Elizabethan and Stuart England*. London: Weidenfeld and Nicholson, 1989.

Deedes, Cecil and H.B. Walters. *The Church Bells of Essex, Their Founders, Inscriptions, Traditions, and Uses*. Aberdeen: W. Jolly and Sons, 1909.

Duffy, Eamon. *The Voices of Morebath: Reformation and Rebellion in an English Village*. New Haven and London: Yale University Press, 2001.

Garrioch, David. 'Sounds of the City: The Soundscapes of Early Modern European Towns.' *Urban History* 30, no. 1 (2003): 5–25.

Gittings, Clare. *Death, Burial and the Individual in Early Modern Britain*. London: Croom Helm, 1984.

Haigh, Christopher. 'The Continuity of Catholicism in the English Reformation.' *Past and Present* 93 (1981): 37–69.

Harding, Vanessa. *The Dead and the Living in Paris and London 1599–1670*. Cambridge: Cambridge University Press, 2002.

James, M.E. 'Obedience and Dissent in Henrician England: The Lincolnshire Rebellion 1536.' *Past and Present* 48, no. 1 (1970): 3–76.

MacKinnon, Dolly. *Earls Colne's Early Modern Landscapes. Aldershot: Ashgate, 2014.*

MacKinnon, Dolly. 'Hearing the English Reformation: Earls Colne, Essex.' In *Hearing Places: Sound, Place, Time and Culture*, edited by R. Bandt, M. Duffy and D. MacKinnon, 255–67. Newcastle: Cambridge Scholars Publishing, 2007.

Morris, Ernest. *The History and Art of Change Ringing*. London: Chapman & Hall, 1931.

Postles, D. '"Ringing out those bells": Death and the Social Order in Early Modern Leicestershire.' *Transactions of the Leicestershire archaeological and Historical Society* 80 (2006): 31–41.

Royal Commission on Historical Monuments of England. *Royal Commission on Historical Monuments (England): Essex*. 4 vols. London: His Majesty's Stationery Office, 1921.

Scott, Joan Wallach. *Gender and the Politics of History*. New York: Columbia University Press, 1988.

Walter, John. '"Abolishing Superstition with Sedition"? The Politics of Popular Iconoclasm in England 1640–1642.' *Past and Present* 183 (2004): 79–124.

Weston, William. *William Weston: The Autobiography of an Elizabethan*. Translated by Philip Caraman. London: Longmans, Green & Co., 1955.

Woodward, J. *Theatre of Death: The Ritual Management of Royal Funerals in Renaissance England*. Woodbridge: The Boydell Press, 1997.

4 The buzz of business

Soundscapes of urbanisation in eighteenth-century London

Markman Ellis

Although London was widely recognised in the early eighteenth century as a city of great size, prosperity and modernity, many commentators also habitually associated the city with excessive noise. In *Amusements Serious and Comical* (1700), the satirist Tom Brown described 'mighty' London as 'this Prodigious and Noisy City,' characterising it as being overwhelmed by 'Crowd, Noise and Perpetual Hurry.' To Brown's narrator, the 'confused Clamours' of the city were enough to 'Stun him, Fright him, and make him Giddy.'[1] These effects – noise, hurry, clamour – are a part of the city's modernity and its prosperity, registered in critical and negative coding. This article is an enquiry into the sonic discourses of London's prosperity and modernity in the period, exploring them in the context of wider negotiations between commerce and formations of civility and politeness. Investigating the connection between civility and noise, it asks whether civility must always be quiet and peaceful. As Hillel Schwartz suggests, towards the end of his very long study *Making Noise*, the urban conditions for civility are not simply engendered by absence, peacefulness and silence.[2] This enquiry proceeds by examining the coffee house and the exchange, two celebrated 'spaces of modernity' in the city (to use Miles Ogborn's phrase).[3] As soundscapes, both these locations produced confusion, clamour and noise, but as this chapter shows, writers in the early eighteenth century, from Ned Ward and Tom Brown to Joseph Addison and Richard Steele, made innovative efforts to describe the distinctive sonic effects of these places. In particular, they observed a characteristic buzz or hum of commerce and assembly that they associated with sociability and civility, although they could not agree upon its dimensions, causes or ramifications.

Ned Ward's ramble satire *The London Spy* offers an extraordinarily detailed poly-sensory account of London life at the turn of the eighteenth century. First published serially in eighteen monthly parts between 1698 and 1700, and then republished whole in 1701, *The London Spy* describes a visit to London by a scholar known only as the Spy, where he undertakes a series of excursions around the city accompanied by a friend. In the third part, the Spy and his friend visit the Royal Exchange, a large and imposing building at the centre of commercial activity in the City of London. Constructed after the fire of 1666 by Edward Jarman, on the site of Thomas Gresham's first exchange built in 1571, it housed the *bourse*, the principal trading floor of the officially sanctioned market in goods, cargoes and shipping. Richard Steele, in *The Spectator* (no. 454), called it 'the Centre of the City, and Centre of the World of Trade.'[4] Ward describes how the Spy and his friend pass through the pillars at the entrance into the interior courtyard:

> We then proceeded and went into the 'Change, turned to the right, and jostled in among a parcel of swarthy buggerantoes (preternatural fornicators, as my friend called them) who would ogle a handsome young man with as much lust as a true-bred English whore-master would gaze upon a beautiful virgin.[5]

Ward's prose style is deliberately vulgar and offensive, elaborately metaphorical and unafraid of misanthropic stereotypes, variously xeno- and homophobic. The ramblers press their way into the crowd of merchants, squeezing past the 'bum-firking Italians,' and on to a 'throng of strait-laced monsters in fur and thrumb-caps' (the Dutch), and further past English Protestants, Spanish, French and Jews, and then Scots, Irish and Americans – each non-equivalent kind assembled in their customary 'walks' on the Exchange floor. Ward has long been recognised for his distinctive attention to visual effects, as here where he notes how the different trades 'ogle' and 'gaze' at each other.[6] But as the Spy enters the courtyard of the Exchange, Ward also notes a distinct auditory event: 'an incessant buzz, like the murmurs of the distant ocean, stood as a diapason to our talk, like the drone to a bag-pipe.'[7]

The Royal Exchange was constructed as a high two-storey quadrangle, perforated by a portico entrance on each of its four sides (Figure 4.1). The central floor was open to the sky, and, during trading hours, thronged with a remarkable congregation of men of commerce, from, as Ward notes, all of Europe and the wider world of British mercantile trade. As a seventeenth-century geography noted:

> In all Cities and Towns of Traffique, there are always found [...] particular places, [...] where Merchants and Tradesmen do assemble and meet at certain hours, and limited times of the day, to confer and treat together, concerning Merchandizing, Shipping, Buying or Selling, and the like; as is seen to be the Royal Exchange of London.[8]

Commercial activity on the Exchange was accomplished by multiple conversations all happening together. The trading floor allowed diverse deals to be made, not by open outcry like an auction, but by direct face-to-face conversation. Ward, like other satirists, made much of this jangle

Figure 4.1 Thomas Bowles, *The inside view of the royal exchange at London*, c.1750, hand-coloured engraving, 28.3 × 42.5 cm, Yale Center for British Art, Paul Mellon Collection.

of languages and discourse, noting the Dutch 'grunting to each other like hogs at their peas,' the Spanish whose 'every sentence ... was performed with a shrug of the shoulders' and the French, 'who talk more with their heads and hands than their tongues.'[9] Out on the street, before entering the Exchange, the Spy is not forewarned of the crowd of merchants inside by any auditory signal: he comes in the gate, and is assailed at the same time by the sight and sound of the assembled throng, all talking together.

Ward's description identifies the sonic experience of the Exchange as quite distinctive in the city. Tom Brown described the Royal Exchange in ways that resonate with these terms:

> That magnificent building there, which stands in the middle of London, is for the accommodation of Lady Trade and her heirs and successors for ever, and is so full of amusements about twelve o'clock every day, that one would think all the world was converted into newsmongers and intelligencers; for such is the first salutation among all mankind that frequent that place. 'What news from Scanderoon and Aleppo?' says the Turkey merchant. 'What price bear currants at Zante? Apes at Tunis? Religion at Rome? Cutting a throat at Naples? Whores at Venice? The cure of a clap at Padua?'[10]

Brown recognises the importance of the Exchange in the commercial and spatial organisation of the city. To his mind, the central activity of the market is not the merchants' trading commodities, but rather their commerce in opinions, gossip and sedition – that medley of intelligence called news. In this quotation, then, Brown satirises how the merchants continuously ask for commercial news from their correspondent ports, by depicting them making an escalating series of requests for ridiculous prices current:

> 'Now,' says I to my Indian, 'is not all this hodge-podge, a pleasant confusion, and a perfect amusement?' The astonished traveller replied, 'Without doubt, the indigested chaos was but an imperfect representation of this congregated huddle. But what most amuses my understanding is to hear 'em speak all languages, and talk of nothing but trucking and bartering, buying and selling, borrowing and lending, paying and receiving; and yet I see nothing they have to dispose of.'

All this discourse amalgamates together in amusement and confusion, albeit of a pleasant variety. The term Brown has for it is 'Hodge-Podge': originally a stew of different meats, and later, any heterogeneous medley or farrago. That culinary metaphor is reinforced by the idea of chaos 'undigested.' The 'congregated huddle' of traders seem to 'speak all languages,' and yet they talk of nothing but trade.

Another description, written a few years later in 1703, in an anonymous satire called *Mirth and Wisdom* (1703), republished a few years later as *Hickelty Pickelty* (1708), also noted the curious sonic cohesion of the Exchange:

> Is the Land's Epitome, or you might call it the Little Isle of Great Britain, did the Waters encompass it. It is more, 'tis the whole World's Map, which you may here discern in its perfectest Motion, jostling and turning. 'Tis a vast heap of Stones, and the confusion of Languages makes it resemble Babel. The noise in it, is like that of Bees; a strange Humming or Buzzing of walking Tongues and Feet; it is a kind of still Roaring, or loud Whisper. It is the great Exchange of all Discourse.[11]

This description of the auditory event of the Exchange builds up a crowded series of sonic images: of jostling, confusion of languages, babel, bees humming and buzzing, still roaring and loud whisper. The sonic spectacle of the Exchange is derived from the variety of languages heard on the trading floor, which together produce a distinct effect, buzzing and humming like bees.

The Exchange is also an important location for *The Spectator*, the essay periodical written and edited by Joseph Addison and Richard Steele in 1711 and 1712. There is considerable literary and ideological distance between the hectic satire of *The London Spy*, and the calm civility of *The Spectator*, although they often inhabit the same spaces. Both, for example, are attracted to the spectacle of the Exchange. Like the Spy, in *The Spectator* no. 69 (19 May 1711) Mr Spectator notices the congregation of traders from around the world on the 'High-Change,' which he describes as 'a great Council, in which all considerable Nations have their Representatives,' listing thereafter inhabitants of Japan, Turks from the Ottoman Empire, Russians, as well as Armenians, Jews, Danes, Swedes and Frenchmen. 'I am infinitely delighted in mixing with these several Ministers of Commerce, as they are distinguished by their different Walks and different Languages.'[12] In *The Spectator* no. 454 (11 August 1712), Steele returns to the Exchange (in an essay describing a day's ramble about the city), observing the 'other men in the Crowds about me [who] were pleased with their Hopes and Bargains.' Mr Spectator goes upstairs in the Exchange and, looking out a window, gets a prospect of the trading floor:

> I went to one of the Windows which opened to the Area below, where all the several Voices lost their Distinction, and rose up in a confused Humming; which created in me a Reflection that could not come into the Mind of any but of one a little too studious; for I said to my self, with a kind of Pun in Thought, *What Nonsense is all the Hurry of this World to those who are above it?*[13]

Like the other writers noticed here, Steele describes the sonic effect of the crowd of merchants as a collective or social sound, a 'confused Humming.'

The distinctive sonic quality of the auditory environment of the Exchange is the buzzing and humming emanating from this congregation of merchants. Ned Ward describes 'an incessant buzz, like the murmurs of the distant ocean, [which] stood as a diapason to our talk, like the drone to a bag-pipe.' His description conjoins five terms: buzz, hum, murmur, diapason and drone. The *Hickelty Pickelty* writer reiterates the language of hum and buzz, as does Steele, and adds the phrases 'still Roaring' and 'loud Whisper.' Each of these warrants further attention. The term 'buzz,' according to the *Oxford English Dictionary*, refers to a 'sibilant hum' such as is made by bees or flies. It is also applied to the 'confused or mingled sound made by a number of people talking or busily occupied' (also *OED*), that is to say, a buzz is created by the fused sound of many voices talking together. Sociolinguists describe this effect as multi-talker babble.[14] Obviously, the sound of a mass of people lends itself to descriptions of the city. Abraham Cowley's poem 'The Wish' (1656) suggests that buzz was typical of city life: contrasting the 'busy world' of the city with the peaceful life of ease available in the country, far from the pride and ambition of the city dweller. Cowley asks:

> Who for it can endure the stings,
> The *Crowd*, and *Buz*, and *Murmurings*,
> Of this great *Hive*, the *City*.[15]

As Cowley suggests, Ward's 'buzz' embeds a bee metaphor. The beehive has been long associated with urban life, from classical Rome to the early modern period.[16] In this period, for example, Bernard Mandeville's *The Fable of the Bees* (1714) uses the hive to launch a satire on the values and manners of a commercial society. The *Hickelty Pickelty* satirist makes explicit the aural connection between this humming and buzzing and the sound of bees.

Linguists argue that the term 'buzz,' like 'hum,' is echoic (that is, it is imitative of a natural sound; what might be called onomatopoeic in literary studies). The presence of these echoic terms in descriptions of the noise of the Exchange is intriguing. An echoic word reflects or imitates the sounds heard by the writer, so that there seems to be a curious connection between the sound event itself, the multi-talker babble of the crowd of merchants and the sounded sound of the writing. Choosing to represent the sound itself of the multi-talker hum of the Exchange floor with the echoic language of buzz or hum, an author encodes this sound in writing, preserving in the onomatopoeia a vestige of the sounded sound, which can be released again by the listener. Without wanting to pursue this too far, echoic or onomatopoeic language seems to work as a kind of recording device: a human one, rather than a mechanical or technological phonograph.[17]

The *Hickelty Pickelty* writer adumbrates the sonic idea of the Exchange's 'strange Humming or Buzzing of walking Tongues and Feet' by introducing two odd phrases, 'still Roaring' and 'loud Whisper.' Both are close to an oxymoron, a figure of speech that juxtaposes elements that appear to be contradictory. The diction of both mingle one term of quiet and calm with one of noise and clamour and suggest that the sonic event must be somewhere between or above a roar (a loud, deep, prolonged sound as made by the sea, a crowd or machinery) and a whisper (a breath sound or low non-resonant quality of voice). Intriguingly, Samuel Taylor Coleridge uses the phrase 'still roaring' to describe the sound of a stream running in an 'o'erwooded, narrow, deep' dell in 'This Lime-Tree Bower my Prison' in 1797, though presumably independently.[18]

Ward's term 'murmur,' of Latin origin, means 'A low continuous sound, such as produced by water, wind,' as is suggested by Ward's metaphor 'the murmurs of the distant ocean.' It also refers to an indistinct utterance of complaint, anger or resentment, which has long associated it with discontent and the city. Murmur is also said by linguists to be of echoic or onomatopoeic origin. Ward's last two terms 'diapason' and 'drone' also go together. A drone is a 'continued deep monotonous sound of humming or buzzing,' such as that made by the continuous bass note of the bagpipe, or the humming of a bee. The term 'drone' keeps the apiarian bee metaphor alive in Ward's sentence but may also be echoic in origin. 'Diapason' is a more curious word, one with specific musical connotations. It derives from a Latin term referring to 'the concord through, or at the interval of, all the notes of the scale' [the interval of an octave], but also figuratively meaning 'Complete concord, harmony, or agreement,' or 'The combination of parts or notes in a harmonious whole, properly in concord.' Ward's term 'diapason' suggests the various voices and languages of the crowded merchants have a kind of harmony or concordance. Together then, these writers suggest, the soundscape of the Royal Exchange can be captured in concurrent and overlapping tropes of bee-like buzz and industry, somewhat natural (oceanic), somewhat in harmony.

Is this soundscape distinctive? In his book *The Acoustic World of Early Modern England*, Bruce R. Smith writes briefly about the soundscape of the Exchange in the early seventeenth century (about a century before my evidence presented here), but does not note anything about a characteristic buzz or hum, although he does note the multiple languages spoken there, making the Exchange 'a genuinely polyglot soundmark.'[19] The buzz of business would then seem to be an important innovation of the eighteenth century, either in sound creation or listening practice. The sound itself could be a new one, perhaps caused by the larger numbers of merchants crowding onto the Exchange in this period, or indeed it may have been produced by the distinctive architecture of Jarman's new Exchange building, erected after the fire of 1666 and opened in 1672. Or it could be that writers are hearing the sound differently, and writing it down using a new echoic language. Or it may point to some wider historical change in the period: for example, to the increased centrality of the Exchange in the financial revolution of the 1690s. Or all three. In any case, the curious harmonies of the buzz of business in the Exchange advertise the close relations imagined in the period between commerce and civility.

Ideas of civility are central to recent accounts of the early modern period. Civility and politeness are explored by numerous texts from the medieval period onwards, notably in the prose and verse forms of the courtesy book, advice book and conduct book. The most influential early treatment of this theme is that of Norbert Elias, for whom the civilising process, 'a restructuring of human personality in the direction of discipline and constraint,' was centred on the royal court. Anna Bryson argues in her book *From Courtesy to Civility* (1998) that the terms civil and civility are keywords for the period, defining the manners and social mores of early modern England. In these accounts of civility, it is noticeable that the concept has a long history, and that most accounts peter out in the late seventeenth century – although, as Larry Klein argues, civility is to some extent subsumed into debates on cognate terms like politeness in the eighteenth century.[20] Nonetheless, in all these accounts, civility marks the long-term movement from the feudal and medieval to an early modern or Enlightenment society. One historian notes 'a transition from a cultural system of lineage, hospitality, fealty and strong communal ties to a nation state of private and possessive individuals and centralised government.'[21]

For the discourse of civility, a central issue posed by eighteenth-century England was the rise of commerce in the aftermath of the financial revolution of the 1690s. Early eighteenth-century moral philosophers repeatedly debated the relationship between civility and commerce, focussing their discussions on notions of commerce and trust, and ideas of civility amongst the middling sort. Both the Royal Exchange and the coffee houses were significant locations in this historical narrative: historiographically speaking, they were identified as amongst the most important arenas in which the new commercial society of England was negotiated. The term civility, and the practice of being civil, Jennifer Richards suggests, was 'synonymous with orderly and accommodating manners, but could also be used to describe metropolitan, commercial, expansionist and self-consciously technological societies.'[22] The Exchange, as has been discussed, was widely perceived as the epicentre of the world of trade, and as such, was a central emblem of both the financial and commercial revolution in the period. The Exchange's harmonic buzz was a distinctive sound-mark of that centrality.

Civility, commerce and the coffee house

Coffee houses were a distinctively modern innovation of the new commercial society in England, and as such were one of the new institutions most closely associated with the new civility. Whereas Ward, Brown, Addison and others heard the distinctive sound-mark of the Exchange as a sign of the harmonies of commerce and civility, the same writers found the noise of the coffee house a less legible event. Coffee is a distinctly modern commodity, first encountered by European travellers in the Levant in the very late sixteenth century. Although some individuals with close contacts amongst the Turkey merchants of the Levant Company drank coffee in the 1630s, the first public coffee house in Christendom opened in London in 1652, close to the Royal Exchange in St Michael's Alley off Cornhill. Coffee houses quickly became characteristic of the distinctive social world of London: there were more than 82 within a decade, and something like 500–600 by the end of the century. The phenomenon of the coffee house was coincident with the broader commercial trends of Restoration and eighteenth-century England. As well as the novelty of coffee itself, the innovation of the coffee house is found in its particular sociability, discursive conversation and connection with the manuscript and print culture of news.[23]

The English coffee house, to contemporary clientele, had a characteristic social space and a distinctive soundscape. This soundscape is no longer accessible, but some vestiges of it are recorded in visual images and textual representations from the period. One particular representation of the coffee house, labelled 'Interior of a Coffee House' (c.1690–1700), elucidates some important aspects of the social and acoustic aspects of the institution in the late seventeenth and early eighteenth centuries (Figure 4.2). The image, a body colour painting by an unidentified artist, indicates how the architecture of the coffee room established the kinds of sociability found in coffee houses: a single large room, up one flight of stairs, with long tables round which the clients sat, talked and read. Around them worked the coffee boys or waiters, bringing the hot beverage in coffee pots from the fire to the customers. None of these activities are particularly noisy, but each aspect of the coffee making, the coffee service and the clients' conversation would have contributed to the acoustic regime of the coffee room. The image suggests various sources of noise beyond speech – it is both a noisy and a smelly picture. The fire, cauldron and pots of the coffee preparation, the waiters and their coffee service, and the papers and pens on the tables: each of these might generate noise, from the rattling of cups and clatter of metal coffee pots, the rustle of paper and the scratch of quill pens. Some of the items depicted in the image stand as metaphors of a sound, and the image as a whole, a depiction of the noisy sonic confusion of the coffee house.

Figure 4.2 Anon, 'Interior of a London Coffee-house,' c.1690–1700, drawing with body colour on paper, 14.7 × 22 cm, © The Trustees of the British Museum.

In a coffee house, drinking coffee was an important and defining activity, but equally or more important was the coffee house's role in sociability and information exchange. The key to this was the custom of the 'common table': when a customer entered the coffee house, he was expected to take the next available free seat at a common table. This produced an unusual degree of social mixing, especially as coffee houses were, to an unusual degree, open and accessible to all kinds and manner of men. In consequence, inside the coffee house, hierarchy was to some important degree erased or elided. Once seated at the common table, a customer was expected to converse with his fellow consumers: that is, to be ready to debate with anyone, on matters of importance, in a manner that was interested and polite, and not too disputatious or incendiary. The inclusive principle of coffee-house conversation was coincident with the principle of open access. However, coffee-house equality and accessibility only extended to men. There were numerous coffee houses in which women worked, but in general women of propriety and good standing were not expected in the coffee room.[24] This had consequences for the sociability of the coffee house, and it must be assumed, its soundscape. The distinctive sociability of both the coffee house and the Exchange depended on the exclusion of women: women were not accorded full participation in the discursive environment of either place, although women are present in subordinated service roles in both. The exclusion of women from these sites of commercial modernity poses interesting questions: to what extent does this gendered exclusion determine the nature of the conversation and debate at that place (the regime of sociability), to what extent does it impact on the soundscape and its characteristic buzz and hum, and to what extent is it embedded in notions of civility?

The nature of coffee-house conversation was central to its role in debates on civility. Conversation and debate in the coffee house has been understood as a correlative of the social architecture of the room, and, according to some, the properties of coffee (the caffeine hypothesis).[25] As Bruce Smith observes, each 'speech community' is located in a distinct 'soundscape': a '*speech* community also constitutes an *acoustic* community. Its identity is maintained not only by what its members say in common, but what they hear in common.'[26] Satires exploit this difference both by describing the distinctive aspects of coffee-house clients' speech in terms of topics and manner of address, and, occasionally, by commenting on the collective soundscape of the coffee room. In the late seventeenth century, numerous small satires ridiculed the social, conversational and informational order of the coffee house – as well as the peculiar black and bitter broth they served. Those aspects which made the coffee house novel and innovative were most often criticised as follies and vices. For their encouragement of debate and conversation on news, politics and religion, amongst an unusually broad and common clientele, coffee houses were accused of being seditious, corrupting and unruly. As an anonymous satirist said in 1673:

> A *Coffee-House* is a *Phanatique Theatre*, a *Hot-House* to flux in for a *clapt understanding*, a *Sympathetical* Cure for the *Gonorrhea* of the Tongue, or a *refin'd Baudy-House*, where *Illegitimate Reports* are got in close *Adultery* between *Lying lips* and *Itching Ears.*[27]

As this suggests, the main contours of the soundscape of the coffee house were established by conversation and debate as much as by the coffee service.

Town satire and sonic experience

Ned Ward's satire, with its intense interests in the idiosyncratic characteristics of verbal expression, allows him an unusual degree of attention to the sonic event of each place he encounters. In his rambles around the 'buzzing metropolis' of London, as he calls it, the Spy makes five visits to coffee houses, in each of which he attentively records aspects of their soundscape. In the first ramble, the Spy and his physician friend step into a coffee house near Aldgate, on the eastern fringe of the City:

> where a parcel of muddling muckworms were as busy as so many rats in an old cheese-loft; some going, some coming, some scribbling, some talking, some drinking, others jangling, and the whole room stinking of tobacco like a Dutch scoot, or a boatswain's cabin.[28]

This description of the busy crush of the smoky coffee room is one of the text's more famous descriptions. Seated in the coffee room, the physician and the Spy identify their fellow inhabitants, noting a projector of madcap schemes, a notorious heiress hunter and a rapacious bookseller. Ward's point seems to be that this is a collection of oddities, an assembly of discrete individuals, rather than a community. The muddle of their writing, talking and smoking is reinforced by the sonic term 'jangling': a term that means 'wrangling, noisy altercation' or the 'dissonant or discordant din of voices' (*OED*). The soundscape of this coffee house would seem to be unlike the harmonious diapason of the Exchange's buzz. But subsequent visits in the book suggest that while coffee houses were always about coffee, smoking, news and conversation, the soundscape of each place could be quite distinctive.

Satire is a critical discourse, in which follies and vices are held up to ridicule, in order to encourage reformation. Ned Ward's form of town satire ridicules typical patterns of thought, manners and discourse associated with types or kinds of people and institutions, animated by carnivalesque and grotesque satirical strategies. In the ninth ramble, the

Spy visits the St James's Coffee House, which he describes as 'the most eminent coffee-house at this end of town.' This coffee house was a large, splendid and renowned institution across the road from the entrance to St James's Palace, and as such, was used extensively by courtiers, gentlemen of the court, army officers and those who wished to curry favour with them. The Spy finds it full of 'a very gaudy crowd of odiferous Tom-essences,' a 'fluttering assembly of snuffing peripatetics.' To the Spy's eye and nose, courtiers are indistinguishable from beaux or dandies, men who give excessive attention to fashion, appearance and social etiquette. The Spy comments that

> though there was an abundance of guests, there was very little to do, for it was as great a rarity to hear anybody call for a dish of Politician's porridge, or any other liquor, as it is to hear a sponger in a company what's to pay, or a beau call for a pipe of tobacco. Their whole exercise was to charge and discharge their nostrils, and keep the curls of their periwigs in their proper order. The clashing of their snuff box lids, in opening and shutting, made more noise than their tongues, and sounded as terrible in my ears as the melancholy ticks of so many death-watches....
>
> They made a humming like so many hornets in a country chimney, not with their talking, but with their whispering over their new minuets and borees, with their hands in their pockets, if freed from their snuff-boxes.[29]

The impoverished conversation in the St James's deadens the characteristic soundscape of the coffee house. In the uncanny calm of this coffee house, then, conversation is conducted in whispers, producing a 'hum' like hornets, accompanied by the snap of snuffbox lids. The absence of conversation and debate in common, Ward implies, is caused by and gives evidence of the rarefied and stultifying social world of the fashionable beau.

But in another coffee house in Aldersgate St, which they visit in Part XII, the Spy finds a cantankerous collection of coffee-house politicians (men who talk of nothing but politics). When the Spy enters the coffee house, he is harangued by one, described as a 'bundle of verbosity,' whose 'tongue,' the Spy says, 'began to flutter about in his mouth like a wild bird trepanned into a cage.' This man is by profession an auctioneer (a professional liar, as the Spy notes) and is a radical Protestant dissenter:

> His voice is as untuneable when he speaks as the screaking of a country sign in a high wind, so that were a blind man to hear him talk he might easily mistake the sound to be the whining of some puppy that wants the dug in his dam's absence.[30]

The sound of the man's discordant voice is captured in the echoic term 'screaking,' denoting and producing a shrill grating sound, further likened to that of a whining puppy. In the Spy's view, the 'babble' of this 'rattle-headed prattle-box' tires 'the ears of the company.'[31] In the Aldersgate coffee house, sectarian politics produces a harsh discordant soundscape, seemingly the opposite of the harmonic buzz of civility and commerce.

Later in Part XII, the Spy enters another coffee house after attending a feast and drinking too much claret, in search of an 'antichristian dose of Muhammadan loblolly,' that is to say some coffee (loblolly: a rustic thick gruel or spoon-meat):

> We went into a great coffee-house by the Temple Gate, where a parcel of grave men were thickening the air with the fumes of their nicotian weed. We sat ourselves down amongst the sage assembly, most of the company, we observed, being as choice of their words as a miser is of his treasure, each seeming as loath to open his mouth as the other his cabinet, which made me think they had either something extraordinary in 'em, that they locked up in their Pythagorean silence, or else that they were a parcel of cunning fools, who having a sense of their infirmities, were unwilling by their talk to discover their ignorance. At last comes in an old newshound, who, in hunting after intelligence was at a great loss, and so enquired of the rest if any straggling news had come that way.[32]

Having criticised the discordant babble of one coffee house, Ward finds the silence of this assembly of natural philosophers as disturbing: neither is desirable. Eventually, the quiet is broken by a 'jolly red-faced old toper,' who initiates a lively debate on church politics, defending the established Anglican church:

> Most of the company agreeing rightly with the old gentleman's sentiments, applauded him very highly for so heartily expressing his affections for the Established Church. This serious speech of the old cavalier's was a key to the hearts of all the rest, who began, after one had opened, like a pack of true beagles at full cry, to hunt down the Church's enemies with all imaginable speed.[33]

The term 'full cry' here refers to the barking of a pack of hounds in full pursuit of their prey. In these examples then, coffee-house conversation was by turns disputatious and entertaining: in the Spy's view, it was also notably communal (there being one central speaker addressing the assembly) and varied (between Dissenter's ranting and the exchange of pleasant anecdotes). As Ward's various descriptions of these conversations and debates suggest, the coffee house was closely associated with political debate, including that amongst the ordinary citizens in common life. This has been much celebrated in coffee-house historiography, establishing a close alliance between the coffee house location and ideas of free speech and public opinion: for example, in twentieth-century discussions of the origin of public opinion by historians such as G.M. Trevelyan, sociologists like Hans Speier and philosophers like Jürgen Habermas.[34] But Ward is unable to see this debate as a harmonic soundscape in which opinions are melded together into one civilised buzz.

Civility and the spectatorial coffee house

Ned Ward insisted the coffee house was not a place of harmony and civility, but rather one of discordance and noise: rather than the diapason and buzz, there is jangle and screaking. In the early eighteenth century, Joseph Addison and Richard Steele's essay periodical *The Spectator* (published daily in 1711–1712) saw the coffee house as one of the four important modern urban locations for their programme of moral reformation. As Steele declared in *The Spectator* no. 10 (Monday, 12 March 1711), the periodical was part of a plan to urbanise philosophy and reform public sociability:

> It was said of Socrates, that he brought Philosophy down from Heaven, to inhabit among Men, and I shall be ambitious to have it said of me, that I have brought Philosophy out of Closets and Libraries, Schools and Colleges, to dwell in Clubs and Assemblies, at Tea-Tables and in Coffee-houses.[35]

These four cultural innovations of the early eighteenth-century city, *The Spectator* argues, are both the site and the model of this project of polite reformation. In *The Spectator* essays, Addison and Steele revise and reform the construction of the coffee house, insisting that it was an integral part of the polite regulation of civil society, and that the soundscape of the coffee room could be tamed.

In *The Spectator* no. 403 (12 June 1712), Mr Spectator considers this 'great City' as the 'aggregate of various Nations distinguished from each other by their respective Customs, Manners and Interests':

> In short, the Inhabitants of St. *James*'s, notwithstanding Language, are a distinct people from those of *Cheapside*, who are likewise removed from those of the *Temple* on the one side, and those of *Smithfield* on the other, by several Climates and Degrees in their way of Thinking and Conversing together.[36]

The cultural coding of these four locations is quite distinct – St James's: court politics; Cheapside: commerce and trade; Temple: lawyers; Smithfield: food. Following this insight, Mr Spectator decides to 'ramble up and down a whole Day together,' listening to the conversation in different coffee houses across the city. He undertakes this day-long coffee-house ramble, 'desirous to learn the Thoughts of our most eminent Politicians on that Occasion.' He follows a news story, a rumour about the death of the French King (Louis XIV, 1638–1715), exploring how the diverse reception of this false news reveals the mentality of each quarter of the city:

> That I might begin as near the Fountain-head as possible, I first of all called in at St *James*'s, where I found the whole outward Room in a Buzz of Politics. The Speculations were but very indifferent towards Door, but grew finer as you advanced the upper end of the Room, and were so very much improved by a Knot of Theorists who sat in the inner Room within the Steams of the Coffee Pot, that I there heard the whole *Spanish* monarchy disposed of, and all the line of *Bourbon* provided for in less than a Quarter of an Hour.[37]

As noted earlier, St James's Coffee-House was a large coffee-house across the street from the entrance to St James's Palace, and as such, attracted a clientele strongly marked by the royal court and its hangers-on, including courtiers, army officers, the great men of state, and those seeking preferment and advantage. In this location, the conversational 'Buzz of Politics' grows 'finer' as it grows closer to the imagined scene of power.

As Mr Spectator journeys across town, visiting the famous coffee houses of each district, he hears different interpretations of the rumour of the king's death: the lawyers and students in the Temple debate the constitutional consequences; around Smithfield, they consider the impact on food prices, especially sprats; and in Aldersgate, nonjuring dissenters vehemently debate an inconsequential point ('whether the late *French* King was most like *Augustus Cæsar*, or *Nero*'). In each coffee house, the debate on the king's death eventually reverts to the clientele's favourite subject, their *idée fixe* or hobby horse. So, in Cheapside, close to the Royal Exchange, the merchants lament their failure to use the knowledge to short the market in stocks and shares. The rumour rumbles on until new news is received, direct from Garraway's Coffee-house, that the French king is alive and well.

Steele's *Spectator* no. 49 (26 April 1711) explores the social soundscape of the coffee house by examining the different kinds of customers found there through the hours of the day.[38] Early in the morning, he finds tradesmen reading newspapers and swapping opinions on the news: later, law students from the Inns of Court dominate the room, some preparing for their day, and others sauntering in their nightgowns, with 'no other purpose but to publish their Laziness.' When 'the Day grows too busy for these gentlemen,' they give place to 'Men who have Business or good Sense in their Faces, and come to the Coffee-house either to transact Affairs or enjoy Conversation.'[39] These are the men Mr Spectator finds most suited to his moral project: they are content 'to be happy and well pleased in a private Condition' – seeking neither political advancement nor the sordid scenes of commerce – while not neglecting 'the Duties and relations of Life.... Of these sort of Men consist the worthier Part of Mankind; of these are all good Fathers, generous Brothers, sincere Friends, and faithful Subjects.'[40] These men of the middling sort, Mr Spectator argues, are the most amenable to his project of moral reformation, which can be accomplished in and through their coffee-house sociability. *The Spectator*, even to near contemporaries, was considered attentive to polite modes of behaviour: Samuel Johnson says that it undertook a role in establishing 'the minuter decencies and inferior duties' and in regulating the 'practice of daily conversation.'[41] That there is a sonic dimension to Steele's argument about coffee-house conversation is presaged by the essay's introductory focus on hearers rather than talkers.

Steele's discourse of convivial civility is dependent on regulation of conversation and the suppression of excessive noise. In his vision, the coffee house becomes the 'Place of Rendezvous to all... thus turned to relish calm and ordinary Life.' Steele proposes an emulative principle as key to the transformation, which he explores through the example of one man, identified by the name 'Eubulus.' He is a kind of ideal citizen amongst men of the middling sort: a rich man who lives modestly, a man of wisdom and influence who holds no political or judicial office, a man who lends money to his friends but does not demand the highest rate of return. Eubulus exerts a moral 'Authority' over all in 'his little Diurnal Audience,' providing a model for their own polite self-reformation. The coffee-house clientele's 'Veneration towards [Eubulus] is so great,' Steele says,

> that when they are in other Company they speak and act after him; are Wise in his Sentences, and are no sooner sat down at their own Tables, but they hope or fear, rejoice or despond as they saw at the Coffee-house. In a word, every Man is *Eubulus* as soon as his Back is turn'd.[42]

Steele's model of Eubulusian regulation of discourse is the process of emulation central to the moral reform of public culture. The example prefigures Norbert Elias's suggestion that emulation is central to the civilising process: 'the compulsion to check one's own behaviour,' through shame and embarrassment, drives the 'structural transformation of society' into its 'new pattern of social relationships.'[43] In Eubulus's coffee house, each man speaks and acts after him, using his example to model behaviour even when not in his presence.

In Steele's essay, the unregulated egalitarianism of the coffee house allows it to become the arena in which men achieve their mannered self-fashioning into the rational and polite residents of the new commercial society. This new civility has a sonic analogue:

> Their Entertainments are derived rather from Reason than Imagination: Which is the cause that there is no Impatience or Instability in their Speech or Action. You see in their Countenances they are at home, and in quiet Possession of the present Instant, as it passes, without desiring to Quicken it by gratifying any Passion, or Prosecuting any new Design.[44]

Eubulus's new coffee house is the site of rational and quiet discussion, without 'Impatience or Instability' in 'Speech and Action.' Coffee-house civility does not raise men's passions but leaves the men in 'quiet Possession of the present Instant.' Conversation in the reformed coffee house will be calm, stable, dispassionate and polite, unlike Ned Ward's carnivalistic and idiosyncratic babble. This reform of coffee-house sociability, in the new civility of its customers, expels the disputatious and rancorous element of coffee-house discussion, which is rendered polite and rational by this civilising process.

But despite these arguments, most satirists and commentators in the eighteenth century continued to find the coffee house a place of carnivalistic noise: bustling with business, buzzing with rumour, humming with gossip. Their evidence suggests that the coffee house was often anything but quiet, polite and business-like, and, moreover, that this disputatious stimulation was a signal source of the customer's interest in going there. Later eighteenth-century coffee-house satires, in verse, essay and prose, celebrate the weird and unpredictable diversity of coffee-house debate, with its babble, disputatiousness, noise and smokiness. A miscellaneous array of figures articulate this counter-polite culture of the coffee house, among which are numbered the gambler and card-shark, the drunkard duellist, the projector (a promoter of madcap schemes), the philosopher and literary critic (given to extreme opinions), the buttonholer (one who literally seizes the observer by the buttonhole, in order to secure undivided attention) and the coffee woman (always ready to gossip and flirt over the coffee-house bar). As a coffee-house critic observed, conversations in coffee houses were likely to distract scholars from their serious work of reflection, 'for who can apply close to a subject with his Head full of the Din of a Coffee-house?'[45]

The enduring concern of most eighteenth-century satirists about coffee-house conversation remained its capacity for immoderate passion and uncontrolled clamour. In the anonymous verse satire *George's Coffee House. A poem*, published in 1761, the subject of coffee-house conversation, debate and gossip, is examined through the poem's gentle imitation of the mock-georgic mode of Gay's *Trivia* (1716). In the poem, the peripatetic muse Clio visits a series of coffee houses, describing their distinctive clientele, and gently satirising their typical discursive regime. So in Jonathan's Coffee-House in 'Change Alley, the principal coffee-house for the market in stocks and bonds, Clio observes a Jewish trader slipping a murky rumour into the room, describing how as 'the Bait went down, the Buz soon general grows.'[46] After visiting Lloyds, the Jerusalem, the Bedford, and Tom's in Russell St, Covent Garden, Clio moves to George's Coffee-House near Temple Bar, on the corner of the Strand and Devereux Court. This coffee house was located close to the lawyers in the Inns of Court, and it was well known for its literary associations (the writers Shenstone, Walpole and Armstrong were all regulars). Clio witnesses the arrival of the mail, bringing news of a reverse in the war in Germany (the context is the Seven Year's War):

> Just now the post deliver'd at the bar,
> His anxious parcels from the Seats of War.

The news provokes much discussion amongst the assembled men, so that:

> Words soon grew high – my Muse had little fears,
> When Physic, Law, and War, were by the ears,
> All talk'd together, equally all big
> From the rose bag to the full bottom'd wig,
> The din still rose, the Widow left her bar,
> And swell'd with fury join'd the clam'rous war.

The coffee woman's intervention is checked by the arrival of even fresher news, declaring a British victory by General Hodgson at Belleisle in Quiberon Bay, at which all together celebrate:

> The theme soon chang'd, joy smil'd in every face,
> And a new lustre glar'd on ev'ry lace.
> So great, so sudden, were the news to all,
> Together, Widow, Will – Will, Widow bawl,
> For Coffee, Punch, Tea, Doctors, Capillaire,
> To drink brave Hodgson, Widow, do you hear?[47]

The whole coffee house calls for drinks and joins in collective patriotic celebration. In this poem, then, coffee-house debate is conflictual and disputatious, driven by impetuous passions, not calm reason – even though in the end, these high feelings can serve a restorative social purpose.

In conclusion, this essay has explored two eighteenth-century discourses addressing the soundscape of urban sociability in eighteenth-century

London. Both the Exchange and the coffee house share a soundscape that contemporaries found remarkable because it was dominated by multi-talker babble. In the early eighteenth century, Ned Ward and others struggled to imagine how these kinds of unstructured and multiple conversations could be considered among the polite practices of modern civility, as they bore no little trace of noise and clamour. The example of the Royal Exchange suggested one solution: to see in the melded social voice of the assembled body of merchants a certain polite harmony, the diapason of commerce, with its characteristics of industrious buzz and hum. Coffee-house debate was more difficult for Ned Ward to accommodate to this discourse, to whom it too often seemed like din and babble. One solution was that proposed by Addison and Steele in *The Spectator*, who, when they imagined how the coffee house might play a role in the reformation of public sociability, proposed that a regulation of conversation might effectively constrain the coffee room to the buzz and hum of politeness and civility.

Notes

1 Tom Brown, *Amusements Serious and Comical*, ed. Arthur Hayward (London: George Routledge, 1927), 11.
2 Hillel Schwartz, *Making Noise: From Babel to the Big Bang & Beyond* (New York: Zone Books, 2011), 841. See also Peter Denney, 'A Taste for Quietness: The Polite Gentleman and the English Landscape Garden,' unpublished paper, 2014.
3 Miles Ogborn, *Spaces of Modernity: London's Geographies 1680–1780* (New York: Guilford Press, 1998).
4 Richard Steele, *The Spectator*, no. 454 (11 August 1712), in Joseph Addison and Richard Steele, *The Spectator*, ed. Donald F. Bond, 5 vols. (Oxford: Clarendon Press, 1965), 4:98–103.
5 Ned Ward, *The London Spy from the Fourth Edition of 1709*, ed. Paul Hyland (East Lansing, MI: Colleagues Press, 1993), 58–59.
6 Peter Briggs notes Ward's 'built-in visual bias' in 'Satiric Strategy in Ned Ward's London Writings,' *Eighteenth-Century Life* 35, no. 2 (2011): 80.
7 Ward, *London Spy*, 58–59.
8 Lewes Roberts, *The Merchants Map of Commerce,* 2nd ed. (London: R. Horn, 1671), 12.
9 Ward, *London Spy*, 59.
10 Brown, *Amusements Serious and Comical*, 21.
11 Anon, *Mirth and Wisdom, In a Miscellany of Different Characters, Relating to Different Persons, and Perswasions* (London: B. Bragg, [1703]), 31–32. New edition called *Hickelty Pickelty: or, a Medley of Characters Adapted to the Age. Relating to Different Persons and Perswasions* (London: [s.n], 1708). It is also worth noting this description is closely plagiarised from John Earle's character of St Paul's Walk in his *Micro-Cosmography* of 1629. Given that this is an entirely different place, which having been destroyed in the fire of 1666 no longer existed in 1708, what does this plagiary or repetition say about the soundscape of the Exchange that a description of it can be recorded in a different time and place?

12 Addison, *The Spectator*, no. 69 (19 May 1711), in Bond, *Spectator*, 1:292–296.
13 Steele, *The Spectator*, no. 454 (11 August 1712), in Bond, *Spectator*, 4:98–103.
14 Ruth Bahr and Elaine Silliman, eds., *Routledge Handbook of Communication Disorders* (Abingdon: Routledge, 2015), 164–65.
15 Abraham Cowley, 'The Wish,' in *The Collected Works of Abraham Cowley*, vol. 2, *Poems (1656), Pt. 1, The Mistress*, ed. Thomas O. Calhoun et al (Newark, NJ: University of Delaware Press, 1993), 2:6–8, 44–46.
16 Cristopher Hollingsworth, *Poetics of the Hive: The Insect Metaphor in Literature* (Iowa City: University of Iowa Press, 2001), 32.
17 Hugh Bredin, 'Onomatopoeia as a Figure and a Linguistic Principle,' *New Literary History* 27, no. 3 (1996): 555–569.
18 Samuel Taylor Coleridge, 'This Lime-Tree Bower My Prison,' in *The Complete Poems*, ed. William Keach (London: Penguin, 1997), 138.
19 Bruce R. Smith, *The Acoustic World of Early Modern England* (Chicago: University of Chicago Press, 1999), 61.
20 Norbert Elias, *The Civilizing Process: The History of Manners*, trans. Edmund Jephcott (Oxford: Blackwell, 1978); Anna Bryson, *From Courtesy to Civility: Changing Codes of Conduct in Early Modern England* (Oxford: Clarendon Press, 1998); Lawrence Klein, *Shaftesbury and the Culture of Politeness: Moral Discourse and Cultural Politics in Early Eighteenth-Century England* (Cambridge: Cambridge University Press, 1994).
21 Jennifer Richards, 'Introduction,' in *Early Modern Civil Discourses*, ed. Jennifer Richards (Basingstoke: Palgrave Macmillan, 2003), 2, paraphrasing Mervyn James, *Family, Lineage, and Civil Society* (Oxford: Clarendon Press, 1974).
22 Richards, 'Introduction,' 1.
23 Markman Ellis, *The Coffee-House: A Cultural History* (London: Weidenfeld and Nicolson, 2004); Lawrence Klein, 'Coffeehouse Civility, 1660–1714: An Aspect of Post-Courtly Culture in England,' *Huntington Library Quarterly* 59, no. 1 (1997), 30–51.
24 Emma Clery, 'Women, Publicity and the Coffee-House Myth,' *Women: A Cultural Review* 2, no. 2 (1991): 168–77; Brian Cowan, 'What Was Masculine About the Public Sphere? Gender and the Coffeehouse Milieu in Post-Restoration England,' *History Workshop Journal* 51 (2001): 127–57; Markman Ellis, 'The Coffee-Women, *The Spectator* and the Public Sphere in the Early-Eighteenth Century,' in *Women and the Public Sphere*, ed. Elizabeth Eger and Charlotte Grant (Cambridge: Cambridge University Press, 2001), 27–52.
25 The spatial or architectural model is pursued by Habermas, and latterly by James van Horn Melton, *The Rise of the Public in Enlightenment Europe* (Cambridge: Cambridge University Press, 2001). For the caffeine hypothesis, see Simon Varey, 'Three Necessary Drugs,' in *1650–1850: Ideas, Aesthetics, and Inquiries in the Early Modern Era*, ed. Kevin Cope (New York: AMS Press, 1998), 4:3–51; Jordan Goodman, 'Excitantia, or How Enlightenment Europe took to Soft Drugs,' in *Consuming Habits: Deconstructing Drugs in History and Anthropology*, ed. Jordan Goodman (London: Routledge, 1995), 126–147.
26 Smith, *Acoustic World*, 46.
27 Anon, *The Character of a Coffee-House, with the Symptomes of a Town-Wit. With Allowance, April 11th 1673* (London: Jonathan Edwin, 1673), 5.
28 Ward, *The London Spy*, 18.
29 Ward, *The London Spy*, 154–55.
30 Ward, *The London Spy*, 209.

31 Ward, *The London Spy*, 201.
32 Ward, *The London Spy*, 219–20.
33 Ward, *The London Spy*, 220–21.
34 See Markman Ellis, 'General Introduction,' *Eighteenth Century Coffee-House Culture*, 4 vols. (London: Pickering & Chatto, 2006), 1:xxx.
35 Addison, *The Spectator*, no. 10 (12 March 1711), in Bond, *Spectator*, 1:44.
36 Addison, *The Spectator*, no. 403 (12 June 1712), in Bond, *Spectator*, 3:506.
37 Addison, *The Spectator*, no. 403 (12 June 1712), in Bond, *Spectator*, 3:507.
38 Steele, *The Spectator*, no. 49 (26 April 1711), in Bond, *Spectator*, 1:208–11.
39 Steele, *The Spectator*, no. 49 (26 April 1711), in Bond, *Spectator*, 1:209–10.
40 Steele, *The Spectator*, no. 49 (26 April 1711), in Bond, *Spectator*, 1:210.
41 Samuel Johnson, 'Addison,' *The Lives of the Poets*, ed. John Middendorf, 3 vols. (New Haven, CT: Yale University Press, 2010).
42 Steele, *The Spectator*, no. 49 (26 April 1711), in Bond, *Spectator*, 1:211.
43 Elias, *Civilising Process*, 66. See also the account of a more aggressive 'controlling gaze' driving moral reform in *The Spectator* by Scott Paul Gordon, 'Voyeuristic Dreams: Mr. Spectator and the Power of Spectacle,' *The Eighteenth Century* 36, no. 1 (1995): 3–23.
44 Steele, *The Spectator*, no. 49 (26 April 1711), in Bond, *Spectator*, 1:211.
45 Roger North, *The Life of the Honourable Sir Dudley North, Knt.*, ed. Montague North (London: John Whiston, 1746), 249.
46 Anon, *George's Coffee House. A Poem* (London: T. Osborne et al, 1761), 4.
47 Anon, *George's Coffee House,* 12–13. Two obscure drinks are mentioned: 'doctors,' presumably some proprietary remedy, and 'capillaire,' an orange-flower flavoured syrup diluted in water.

Bibliography

Primary sources

Addison, Joseph and Richard Steele. *The Spectator* (1711–1714). Edited by Donald F. Bond. 5 vols. Oxford: Clarendon Press, 1965.

Anon. *The Character of a Coffee-House, with the Symptomes of a Town-Wit. With Allowance, April 11th 1673*. London: Jonathan Edwin, 1673.

Anon. *George's Coffee House. A Poem*. London: T. Osborne et al., 1761.

Anon. *Mirth and Wisdom, in a Miscellany of Different Characters, Relating to Different Persons and Perswasions*. London: B. Bragg, 1703.

Anon. *Hickelty Pickelty; or, a Medley of Characters Adapted to the Age. Relating to different Persons and Perswasions*. London: [s.n.], 1708.

Brown, Thomas. *Amusements, Serious and Comical, and Other Works*. Edited by Arthur Hayward. London: G. Routledge, 1927.

Coleridge, Samuel Taylor. *The Complete Poems*. Edited by William Keach. London: Penguin, 1997.

Cowley, Abraham. *The Collected Works of Abraham Cowley. Vol. 2, Poems (1656). Pt. 1, The Mistress*. Edited by Thomas O. Calhoun, Laurence Heyworth, J. Robert King. Newark, NJ: University of Delaware Press, 1993.

Johnson, Samuel. *The Lives of the Poets*. Edited by John Middendorf. 3 vols. New Haven, CT: Yale University Press, 2010.

North, Roger. *The Life of the Honourable Sir Dudley North,... and of Dr. John North, Master of Trinity College,... Cambridge, Etc.* Edited by Montagu North. London: J. Whiston, 1744.

Roberts, Lewes. *The Merchants Map of Commerce: Wherein the Universal Manner and Matter of Trade Is Compendiously Handled*. 2nd Ed. London: R. Horn, 1671.

Ward, Edward. *The London Spy from the Fourth Edition of 1709*. Edited by Paul Hyland. East Lansing: Colleagues Press, 1993.

Secondary sources

Bahr, Ruth Huntley and Elaine R. Silliman, eds. *Routledge Handbook of Communication Disorders*. Abingdon: Routledge, 2015.

Bredin, Hugh. 'Onomatopoeia as a Figure and a Linguistic Principle.' *New Literary History* 27, no. 3 (1996): 555–569.

Briggs, Peter M. 'Satiric Strategy in Ned Ward's London Writings.' *Eighteenth-Century Life* 35, no. 2 (2011): 76–101.

Bryson, Anna. *From Courtesy to Civility: Changing Codes of Conduct in Early Modern England*. Oxford: Clarendon Press, 1998.

Clery, Emma J. 'Women, Publicity and the Coffee-house Myth.' *Women: A Cultural Review* 2, no. 2 (1991): 168–77.

Cowan, Brian. 'What Was Masculine About the Public Sphere? Gender and the Coffeehouse Milieu in Post-Restoration England.' *History Workshop Journal* 51 (2001): 127–57.

Denney, Peter. 'A Taste for Quietness: The Polite Gentleman and the English Landscape Garden.' Unpublished Paper, 2014.

Elias, Norbert. *The Civilizing Process: The History of Manners*. Translated by Edmund Jephcott. Oxford: Blackwell, 1978.

Ellis, Markman. *The Coffee-House: A Cultural History*. London: Weidenfeld & Nicolson, 2004.

Ellis, Markman. 'The Coffee-Women, *The Spectator* and the Public Sphere in the Early-Eighteenth Century.' In *Women, Writing and the Public Sphere, 1700–1830*, edited by Elizabeth Eger, Charlotte Grant, Penny Warburton, and Clíona Ó Gallchoir, 27–52. Cambridge: Cambridge University Press, 2001.

Ellis, Markman, gen. ed. *Eighteenth-Century Coffee-House Culture*. 4 vols. London: Pickering & Chatto, 2006.

Goodman, Jordan. 'Excitantia, or How Enlightenment Europe took to Soft Drugs.' In *Consuming Habits: Drugs in History and Anthropology*, edited by Paul E Lovejoy, Andrew Sherratt, and Jordan Goodman, 126–147. London: Routledge, 1995.

Gordon, Scott Paul. 'Voyeuristic Dreams: Mr Spectator and the Power of Spectacle.' *The Eighteenth Century* 36, no. 1 (1995): 3–23.

Hollingsworth, Cristopher. *Poetics of the Hive: The Insect Metaphor in Literature*. Iowa City: University of Iowa City, 2001.

James, Mervyn. *Family, Lineage and Civil Society. A Study of Society, Politics and Mentality in the Durham Region, 1500–1640*. Oxford: Clarendon Press, 1974.

Klein, Lawrence E. 'Coffeehouse Civility, 1660–1714: An Aspect of Post-Courtly Culture in England.' *Huntington Library Quarterly* 59, no. 1 (1996): 31–51.

Klein, Lawrence E. *Shaftesbury and the Culture of Politeness: Moral Discourse and Cultural Politics in Early Eighteenth-Century England*. Cambridge: Cambridge University Press, 1994.

Melton, James Van Horn. *The Rise of the Public in Enlightenment Europe.* Cambridge: Cambridge University Press, 2001.

Ogborn, Miles. *Spaces of Modernity: London's Geographies, 1680–1780.* New York: Guilford Press, 1998.

Richards, Jennifer. 'Introduction.' In *Early Modern Civil Discourses,* edited by Jennifer Richards, 1–18. Basingstoke: Palgrave Macmillan, 2003.

Schwartz, Hillel. *Making Noise: From Babel to the Big Bang & Beyond.* New York: Zone Books, 2011.

Smith, Bruce R. *The Acoustic World of Early Modern England: Attending to the O-Factor.* Chicago: University of Chicago Press, 1999.

Varey, Simon. 'Three Necessary Drugs.' In *1650–1850: Ideas, Aesthetics, and Inquiries in the Early Modern Era*, edited by Kevin Cope, 4:3–51. New York: AMS Press, 1998.

5 Civil noise and its discontents

David Ellison

Recent scholarship in eighteenth-century British history has done much to complicate the apparently stark opposition between polite and impolite cultures. Politeness emerges from these accounts as a much roomier, cross-class category capable of embracing, in Paul Langford's words, 'a great variety of values, attitudes and ideas,' including behaviour that would appear to be its opposite.[1] It is from within these efforts to challenge self-evidently antithetical terms that this chapter finds its subject: civil noise; noise acknowledged as something in excess of articulate, intended or desired sound, yet not presumptively excluded on those grounds. For us, civil noise is likely encountered in provisions protecting the sonically besieged, as in civil noise abatement or civil noise ordinances. Such instruments assume that civility perishes under noise and that legal curtailment is restorative. To some extent, the emergence of the discourse of politeness initiated, or at least consolidated, this link between noise and incivility, leading to numerous attempts to regulate the acoustic environment of urban life.[2] Less noticed in both studies of sound and civility, however, is that, for extended periods in the eighteenth and early nineteenth centuries, noise was also heard from within the soundscape of English politeness, not countering it.

My intention in this essay is not to broadly survey the shifting state of English manners, but rather to narrowly examine an early and ultimately protracted example of the decoupling of noise from civility among visitors to so-called 'shew-houses.' I will make three intertwining claims: first, that the great house is one of the locations where the civil noise of a new class of visitors is felt to disrupt the traditional sonic order; second, that early museums addressed the problem of civil noise as part of a set of questions around the function and purpose of public institutions; and third, that the separation of civility from noise shaped the architecture of nineteenth-century middle-class domesticity. What follows is a series of fictional and non-fictional encounters with the expression of noisy civility and the conditions under which it might be disciplined.

What did civil noise sound like? A remnant echo of sorts may be heard in James Beresford's compendious work, *The Miseries of Human Life* (1806). *The Miseries* proceeds as a series of alternating monologues from Timothy Testy, a coarse, city-dwelling misanthropist, and Sir Samuel Sensitive, country gentleman, all-suffering devotee of sensibility and fleeting domestic comfort's mourner-in-chief. While each attempt to convince the other that he alone has suffered the most, cumulatively they present over 300 'groans' grouped by their originating sources in the miseries of dining, reading, writing, travel, fashion, the country, the town and the home.

Much of what is negatively encountered in *The Miseries* would be all too familiar to casual readers of Smollett, Sterne or Fielding: unstoppable bores, incontinent servants, poisonous cooks, cheating tailors and the manufacturers of defective flues. Added to these durable sources of unhappiness are those that originate from within Testy and Sensitive themselves: gout, embarrassing turns of forgetfulness and a nearly inexhaustible supply of evening-ruining *faux pas*. The satire of *The Miseries* casts a wide net, drawing in a host of middling English agonies measured along the length of their put-upon bodies: 'What, my poor Sir,' asks Sensitive, 'are the senses but five yawning inlets to hourly and momentary molestations?'[3] Not surprisingly, many of these insults prove sonic in origin; with both Sensitive and Testy finding occasion to complain of strident bells, deafening conversational settings, incursive street noise and tuneless footmen given to song. Some disturbances are heard to cluster around sociable sites like the coffee house, theatre or concert hall, where noise specifically interferes with the capacity to direct attention: 'At a concert – as you are preparing to listen to one of Bartleman's best songs, being suddenly environed by a crew of savages, whose laughter and gabble are all that you are allowed to hear.'[4]

When he calls out a portion of the concert audience as savages, Sensitive applies a term more at home in the denigration of plebeian cultures. Certainly, this is how he uses it in the 'Miseries of London,' when he complains of the brute, wall-penetrating force of 'that savage jargon of yells, brays, and screams' coming from 'dustmen, beggars, muffin-mongers, knife grinder, and news carriers.'[5] Similarly, Testy uses it when he describes the 'savage jollity' of popular amusements such as 'bull-baiting, a boxing match, an execution, &c &c.'[6] Sensitive may wish to shame the noisy concertgoers surrounding him, but by some accounts, these savages were not a distinct cohort, but rather the entire audience. Horace Walpole makes an earlier, related observation about the London opera audience in a letter to Sir Horace Mann in which he complained of 'those who don't love music, love noise and party, and will any night give half-a-guinea for the liberty of hissing.'[7] Walpole's point is that music is incidental to the more pressing demands of things like faction (party). It is this audience engaged with the supposedly wrong object that emerges in Lydia Goehr's *The Imaginary Museum of Musical Works* (1992). Goehr argues that English and European concertgoers attended performances that were effectively extramusical – that is, the music served as 'background affairs ... to serious or frivolous activities' pursued across box, pit and gallery.[8] William Weber takes issue with Goehr's characterisation because she judges the eighteenth-century audience – and finds them wanting – by post-Romantic standards of serious, silent and sustained attention across the musical work. He suggests instead that opera audience behaviours and listening practices are better understood within a context of what he terms the 'sociable aesthetic.'[9] Weber concedes that audiences both chatted loudly and moved about, but this was mostly in gaps opened between those parts of the performance that were most highly valued, such as the aria. From a post-Romantic vantage point that 'distrusts any fusion between music and mundane social activities,' this kind of disorderly behaviour is unconscionable.[10] But, for Weber, Goehr's judgement fails to properly acknowledge the complex demands placed on the audience to actively participate in the performance as social event.[11] Although he does not refer to *The Miseries* directly, Weber would likely view Sensitive's complaint against the savages as evidence of the existence of a listening principle within the totality of the sociable aesthetic itself, but it might be better understood as a dissenting argument urging the suspension of otherwise permissible Georgian noise under conditions of cultural appreciation.

In the eighth groan, Sensitive again bridles in the midst of an unruly audience, this time at the restrictions their unregulated behaviour places on the free expression of his emotions:

> Your feelings put to the rack throughout the most moving scenes of a deep tragedy, by a riotous rascal in the upper gallery, who will not, for a moment, suffer his neighbours to cry in peace – while you are perpetually tantalized with neglected proposals from the tender-hearted part of the audience, 'throw him over.'[12]

Sensitive's complaint acoustically distinguishes three rival domains of speech filling the space from stage to upper gallery: dramatic performance, disruptive mockery above and countering hostility below. These combine to form the unruly (but otherwise unremarkably civil) noisescape of the theatre from which Sensitive seeks respite or, rather, a bubble of amenable silence in which to weep.

His final groan on this subject – and after three related complaints of this kind, we may call it a subject – moves beyond the performance halls and theatres to a 'shew-house':

> In seeing what is called 'a Shew-house,' – keeping pace, whether you will or not, through all the rooms, with another party, (Hottentots,) – by which means, besides having your privacy destroyed, you cannot hear (or cannot understand), what is said by one guide, for the continual counter-gabble of the other.[13]

The savages plaguing performance hall and theatre make a hyperbolic return in a racially specific form to renew their noisier assault on whatever it is that Sensitive aspires to learn about 'all the rooms.' His peculiar response to their apparent intrusion – that his privacy has been destroyed – implies an alternative, slightly ascetic model of cultural experience in which the sonically insulated auditor needs to hang on the guide's every expert word (perhaps anticipating the advent of the exhibition audio guide). Although Sensitive's private encounter was clearly spoiled, his description of the 'Hottentot's' interfering noise as 'counter-gabble' complicates matters somewhat. The objectionable counter-gabble can find an opposing force nowhere else except in the house guide's gabble, fixing the hapless Sensitive between competing forms of mutually worthless speech. Once again, the totality of sociable and civil noise blocks Sensitive's aesthetic pleasure, self-expression and desire for self-improvement.

Placing the problem of noisy fellow visitors to one side for a moment, the unworthiness of the guide's speech was a commonly remarked-upon feature of the great house visit. On his arrival at Belvoir Castle on 7 June 1789, John Byng sought admission from a housekeeper whose 'very drunken, dawdling appearance' prefigured her 'numberless' mistakes including identifying a painting of the Duke of Buckingham as 'that villain Felton; finely confusing the Murder'd with the Murderer!'[14] The complaints of visitors to 'shew-houses' ranged from the guide's insolence and ignorance[15] to their use of unintelligible jargon – a telling grievance from newer visitors alienated by tours shaped by antiquarian preoccupations.[16] Such experiential faultiness divided the newer, pleasure-seeking visitors from the expectations of more scholarly types. One 'disgusted connoisseur' deplored the way visiting parties were 'driven' through the apartments of Hampton Court:

> by the deputy-housekeeper or one of her housemaids, who pointed out the pictures with a long stick, calling out, in a loud voice, at the same time, the names of the subjects and their painters to the awe stricken company – a procedure that allowed of little opportunity for studying or enjoying them.[17]

In *The British Tourists* (1798–1800), William Mavor's harrumphing dissatisfaction with his tour extended beyond the housekeeper to a host of reputation-damaging staff:

> We have more than once had occasion to remark, that the impression visitors receive from the view of a place, is as frequently conveyed by the mode in which it is shewn, as by its native beauties or defects. A gentleman meets with a surly porter, an avaricious housekeeper, or a begging, insolent gardener; he is at once disgusted; and retorts, very unjustly, the faults of the servant on the owner, or his seat.[18]

Cumulatively, these examples suggest a problem with the house tour where narrative expertise was delegated to those with an unwarranted claim to possess it. It is also the case that there were servants who admirably discharged their responsibility as guides. Visitors to Kedleston Hall in Derbyshire, for example, praised Mrs Garnett's capacity for 'uncommon politeness and attention.'[19] Boswell, who visited the house with Johnson in 1777, paid equal tribute to her seemliness and sonic authority. She was 'a well-drest elderly housekeeper, a most distinct articulator.'[20]

But before Boswell and Johnson could appreciate Mrs Garnett's fabled diction, they, like most visitors to great houses, were obliged to apply for access in writing or, gallingly, present themselves at the gate for an impromptu social assessment based on the quality of their dress, number of servants and mode of transport. With the rise of polite tourism towards the end of the eighteenth century, these methods gave way to new and more restrictive controls such as advance ticketing, sometimes with a list of house rules attached.[21] Walpole, who had no qualms describing visitors to Strawberry Hill as a 'plague,' restricted access to the house to one party of four per day, with the additional proviso that children not be admitted.[22] Besides quashing the – admittedly surprising – prospect of Strawberry Hill's family-friendliness, the ruling against children pre-emptively installs prohibited visitor behaviours in their unruly, sensation-seeking bodies. What these behaviours might have been is vividly demonstrated in Walpole's correspondence. In a 1793 letter to Mary Berry, he writes that

> Two companies had been to see my house last week, and one of the parties, as vulgar people always see with the ends of their fingers, had broken off the end of my invaluable eagle's bill, and to conceal their mischief, had pocketed the piece[23]

Walpole noted other, related incidents experienced by those who, like him, understood 'what a grievance it is to have a house worth being seen.'[24] In a letter to the antiquarian William Cole, Walpole recalls a visit to Cannons, James Brydges's seat near Middlesex, where he saw a 'beautiful table of Oriental alabaster that had been split in two by a buck in boots jumping up backwards to sit upon it.'[25] In a related vein, he notes that Phillip Southcote's famed garden at Wooburn Farm is closed because 'the savages who came as connoisseurs scribbled a thousand brutalities, in the buildings, upon his religion.'[26]

Such sensory overreach is understood in terms of rough or sectarian behaviour. Whatever panicky remorse may have consumed the visitor who snapped the invaluable eagle's bill is not a topic Walpole explores. That is, he only records the visit's brutal aftermath as symptoms of character, his letters never registering, much less imagining, what might motivate the visitor's touch or how they might choose to explain their actions. Fortunately, some traces of what we might think of as the visitor's perspective are recounted in a skirmish between the underkeeper (curator) of the Oxford Ashmolean and a very hands-on museum-goer:

> She desired me to take the Glass from off several of the Drawers, which I was somewhat unwilling to do, lest anything be lost by that means; which she perceiving she told me that I was not quite so civil as might be; that the last time she had seen the Museum ... she had handled and examin'd the Curiosities in the Cabinet as long as she pleas'd.[27]

Of course, the museum – an institution I will turn to shortly – is not identical to the great house, but, as this example suggests, they are both burdened by the visitor's apparent incapacity to keep their hands off the displayed objects – a claim countered in the Ashmolean visitor's rejection of the museum's newly restrictive, object-hoarding order. Her view that the underkeeper's refusal amounts to incivility challenges his effort to cast her demands in a boorish and uncivil light. In other words, both visitor and curator make appeal to civility by way of reining in the other, suggesting a category still surprisingly roomy enough to warrant action on opposing sides of a dispute. In this instance, as Constance Classen notes, the visitor's civil curiosity-rights prevailed and the curator handed over some of the requested items, 'that she might inspect them more narrowly.'[28] For my purposes, the institutional uncertainty around such indelicacies of touch stand in for the cognate problem of barely permissible civil noise as a phenomenon similarly on the verge of disciplinary restriction.[29]

To properly measure the volume of sonic disturbance in the great house, we need to take a step back, as it were, to think in terms of the larger acoustic environment in which civil noise propagated in locations like the long gallery. Visitors to great houses with a keen eye for paintings would have looked forward to touring the long or state gallery. Coming into vogue in the second half of the sixteenth century, the gallery was, in the words of biographer Roger North, a 'room for no other use but pastime and health, so far as the gentle moving usuall within the walls of an house may concerne it.'[30] A space for mobile recreation, it could either be 'taken into the parade of the house' – that is, as quasi-public – or 'kept for the private divertion of the master.'[31] In either case, it warranted decoration mostly in the form of paintings, typically portraits.[32] When Celia Fiennes visited Euston Hall in the late seventeenth century, she was struck by the portraits of the 'Royal Family from K Henry the 7th by the Scottish race his eldest daughter down to the present King William and his Queen Mary.' These were arranged along one side of the long gallery. Along the other were to be found 'foreign princes from the Emperor of Moroccoe the Northern and Southern

Princes and the Emperour of Germany.'[33] A gallery such as Grafton's could signal loyalty to the crown, while also registering more cosmopolitan interests in royal courts abroad. Polite visitors may or may not have understood, or even wished to engage with, the various claims such an arrangement might communicate.[34] In 1767, Weeden Butler, cleric and editor, dramatised a day at the British Museum accompanied by his friend's daughters. The resulting manuscript features young women crying out, 'very pretty! Extremely fine! Excessively curious! Wonderfully disposed! Amazingly elegant! Strikingly beautiful, indeed!'[35] Although Butler's characters are not visiting a great house, their response to the paintings finds its comic premise in the – slightly weary – recognition of the polite visitor's formalised response.[36] Clearly, not all paintings were viewed this way and polite tourism doubtless encompassed many different intensities of looking, including something approaching antiquarian absorption. Butler's account is useful, though, to the extent that it models liberality – the visitor's implied eye dictating progress through the gallery – mixed with something else: formalised speech as sound. The young women's exuberant praise affirms the value of the esteemed object, but it also marks the terminal point of their attention while declaring a readiness to move along towards whatever comes next. Helen Rees Leahy rightly describes the long gallery as a location where 'walking and looking were explicitly conjoined,' but sound is also a part of this assemblage.[37] The sounds associated with movement through the hall would alternate between footsteps, the guide's explanatory speech and the polite tourist's conventionalised appreciation: striking, fine, wonderful. This, admittedly ideal, transaction cleaves closest to the monovocality that Samuel Sensitive expects and grieves in its absence.

Monovocality, it would appear, is particularly well suited to the distinctive and socially aggrandising acoustics of the hall; a large, relatively unimpeded volume with a frequently noted tendency to echo – unsurprising given the larger purpose of the house to amplify name and reputation. Consider this anonymous account of a visit to Hampton Court:

> These princely halls … have come to be almost as silent as their dead master's tomb. They have nothing to echo back but the hurried footstep of a single domestic, who passes through them daily, to wipe away the dust of their untrodden floors, only it may collect there again; or the unintelligible jargon of a superannuated dependant, as he describes to a few straggling visitors (without looking at them either) the objects of art that have been deposited in them, like treasure in a tomb.[38]

For Ernest Law, author of a late-Victorian multivolume history of Hampton Court, the quote furnished evidence of the palace's general decline under George IV. That is, the hall's acoustic characteristics were recruited for the grimmer purpose of ironic reflection; the echoes amplify the *wrong* source. The housekeeper and yet another gabbling dependant occupy acoustic space we are to understand as properly reserved for their master's voice.

The hall's tendency to amplify or modify the circuit of the voice proved an enduring object of interest for nineteenth-century acousticians thinking about the effective propagation of sound in public buildings. It is here where some of the sonic deficits associated with the hall are noted in a technical light. T. Roger Smith's *A Rudimentary Treatise on the Acoustics of Public Buildings* (1861) ruefully acknowledged the difficulty of describing the effect of complex space on sound:

> the truth is, that the moment a sound becomes surrounded by the multiplied conditions that affect it in a building – the disturbance, the resonance, or the echo that it may meet with – it is exposed to so many conflicting influences that it becomes very difficult for the man of science to do more than state very generally what results the conditions under which it is placed may be reasonably expected to induce.[39]

Among those conditions, Smith warns against undamped spaciousness creating an 'injurious excess of sound.'[40] Apartments of 'bad proportion … such as are disproportionately lofty, and have plain unbroken walls, seem very subject to reverberation.'[41] Such spaces are 'liable to be unduly sonorous.'[42]

Although Smith is silent on the social implications of lofty disproportion, he does identify some of the sources of acoustic indistinction in the grander features that both dignify public building and incidentally trap volumes of noisy air. He lays out the problem by quoting a translated passage from Theodore Lachez's *Optique et Acoustique* (1848):

> The coffers … with which ceilings are decorated, vaults, ornamental niches, the reveal of doors, and other similar recesses, all afford little cubes of air which vibrate each one with its own appropriate vibrations, like a kind of organ-pipe, under the excitement of the undulations in the building, and add to the fatiguing resonance of halls by augmenting the sum of confused noises.[43]

Although the introduction of classical, ecclesiastical and aristocratic design elements furnished some grandeur, they potentially muddled the prospect of clear communication. The solutions Smith recommends are straightforward. At Exeter Hall, for example, he briskly notes that a plain ceiling replaced a coffered one 'with success.'[44] Again, there is no suggestion that better results stem from simpler, more demotic form, only that 'bad proportion,' 'ill disposed forms' and 'unshapeliness' are 'possibly the cause of failures which cannot be attributed to any influence that has yet been described.'[45] Buildings that make too much clamour out of speech or song may find their remedy in the technical disciplining of their form by taking something away like prestige coffering. Cumulatively, we see two sides of a combined effort to make civil buildings out of noisy ones. Halls limited their acoustically reflective surfaces, much as the great houses constrained the acoustic conduct of their visitors. The early museums made a related contribution to this project by identifying specific sonic behaviours as uncivil.

As Helen Rees Leahy illustrates in *Museum Bodies: The Politics and Practices of Visiting and Viewing*, the question of who should (and should not) be admitted to locations such as the British Museum and the Royal Academy was a matter of debate throughout the reign of George III.[46] Curators principally occupied with protecting their collections worried about visitors motivated by things other than a desire to see art, such as keeping warm and dry, for example, while those critics and politicians committed to a more inclusive cultural sphere looked to the museum as a space of aesthetic education.[47] Both of these camps, however, scrutinised spectators for evidence of appropriate behaviour – a process that contributed to the visualisation and codification of incivility.

Even before an institution like the British Museum was opened to the public, a committee was established to consider rules governing the visit. The resulting draft memorandum identified 'persons of mean and low Degree' for exclusion, but it also forbade 'Rude or ill Behaviour'; a vaporous category permitting discipline over a much broader range of conduct.[48]

Among those feeling the practical force of this new regime was the historian of Birmingham, William Hutton, who visited the Museum on 7 December 1784. He had applied for an advance ticket – a technique borrowed from the great houses – and, on entry, joined a group of ten on a guided tour:

> We began to move pretty fast, when I asked with some surprise, whether there were none to inform us what the curiosities were as we went on? A tall genteel young man in person, who seemed to be our conductor, replied with some warmth, 'What? Would you have me tell you every thing in the Museum? How is it possible? Besides, are not the names written upon many of them?' I was too much humbled by this reply to utter another word. The company seemed influenced: they made haste and were silent. No voice was heard but in whispers …

When our leader opens the door of another apartment, the silent language of that action is, *come along*.[49]

Intensely disappointed with the experience of the Museum, Hutton likened himself to Tantalus: 'in the midst of a rich entertainment, consisting of ten thousand rarities, but … I could not taste one.'[50] Far from learning anything useful about the collection, Hutton found himself the object of instruction, in compliance, speed and, above all, in the silence and near-silence of civil comportment. Hutton's stifled experience is contradicted in accounts from more permissive institutions like Somerset House, the Academy of Art or even the British Museum under later, more liberal administrations. Henry Thomas Alken's illustrated novel *Real Life in London. Or, the Rambles and Adventures of Bob Tallyho, Esq. and His Cousin, the Hon. Tom Dashall, through the Metropolis* (1821) includes a section devoted to a visit to the British Museum that treats it much like any other site of raucous Georgian sociability; an occasion for gossip, flirtation and satirical reflection:

> the reconnoitering eye of Dashall observed a trio, from whence he anticipated considerable amusement. It was a family triumvirate, formed of an old Bachelor, whose cent per cent ideas predominated over every other, wheresoever situated or howsoever employed; his maiden Sister, prim, starch and antiquated; and their hopeful Nephew, a complete coxcomb, that is, in full possession of the requisite concomitants – ignorance and impudence, and arrayed in the first style of the most exquisite dandyism.[51]

Each of these characters speaks directly to the subject of their interests. For the sister, this was the hope of seeing, as she expressed it, the 'He-gipsyian munhuments, kivered with kerry-glee-fix.'[52] Her brother's concern was to ascertain 'whether, independent of outlandish baubles, gimcracks and gewgaws, there was anything of substantiality with which to enhance the per contra side in the Account Current between the British Museum and the Public!'[53] Nephew Jasper – the dandy – explains the surprisingly Celtic origins of an exhibited sarcophagus, or 'Sark o' Fegus,' a winding sheet that borrowed its name from chemises worn by 'the daughters of O'Fegus, a highland chieftain.'

It is the character's outlandish speech – founded in malapropism and fixation – that comically warrants, and diverts, Dashall and Tom's attention. The accompanying illustration (Figure 5.1) concedes the institution as a provisional stage for the expression of eccentric utterance, pushing or displacing the vaunted objects of the museum to the background and then summarily cutting off their full length. Grouped downstage, the speakers theatrically project ludicrous observations – there is no space or prospect for Samuel Sensitive-like withdrawal into a private exchange with a guide or even quieter appreciation. The comic charge here, such as it is, relies on the triumvirate's resounding cluelessness; they do not or cannot grasp the social risk that follows their exposure to satirical ridicule. Alken does not elaborate or frame the inevitable consequences as a sort of moral: the cancelled engagement, the cold shoulder at the club, the vicar's slight. There are no lessons to be taught along those lines, as this is not that kind of book, but the message is clear enough: the noisy, confident, civility of the museum triumvirate, at least in this location, is terminally marked.

Collectively, the efforts of architects, masters, administrators, acousticians and satirists begin to foreclose on the locations in which noisy civility may be expressed. In order to trace one final and very effective example of silencing, we must turn, briefly and by way of conclusion, to the domestic fortunes of the corridor.

I will not rehearse the early modern European origins of the corridor here in favour of simply observing the fact of its eventual triumph in nineteenth-century Britain. The English architectural historian Robin Evans bundles the corridor with a number of other features that herald the advent of modernity, effectively displacing an older model of freely circulating sociality (idealised in the Italian villa) with organisational forms that privileged efficiency, circulation and, above all, privacy. The corridor contributes to an architecture responsible for 'obliterating vast areas of social experience – reducing noise transmission, differentiating movement patterns., suppressing smells … veiling embarrassment, closeting indecency.'[54] In Evans's account, the corridor embeds domestic privacy through sensory suppression. He draws on an illustrative example of Robert Kerr, author of *The Gentleman's House*, who viewed architecture

Figure 5.1 Thomas Henry Alken, *British Museum. Tom and Bob in Search of the Antique*, 1821, hand-coloured etching and aquatint, 12.9 × 20.2 cm, © The Trustees of the British Museum.

as an exercise in separating gentlemen from the irritations of daily life which Evans lists as 'the mixing of servants and family, the racket of children, and the prattle of women.'[55] The corridor managed the infiltrative and intrusive flow of noise, separating labour zones from spaces of retreat, as well as preserving the family from accidental contact with servants obliged to travel through dedicated passageways. Such a house spatially enshrines – and generalises – the principle of separating the production of noise from the experience of civility as irreconcilable opposites.

Notes

1 Paul Langford, 'The Uses of Eighteenth-Century Politeness,' *Transactions of the Royal Historical Society* 12 (2002): 326. See also, for example, Jenny Davidson, *Hypocrisy and the Politics of Politeness: Manners and Morals from Locke to Austen* (Cambridge: Cambridge University Press, 2004); Jon Mee, *Conversible Worlds: Literature, Contention, and Community, 1762–1830* (Oxford: Oxford University Press, 2011).
2 Emily Cockayne, *Hubbub: Filth, Noise and Stench in England, 1600–1770* (New Haven: Yale University Press, 2007), 121–27.
3 James Beresford, *The Miseries of Human Life* (London: W. Bulmar, 1806), 13.
4 Beresford, *Miseries*, 85.
5 Beresford, *Miseries*, 63.
6 Beresford, *Miseries*, 37.
7 Horace Walpole, *The Letters of Horace Walpole, Earl of Orford*, ed. John Wright, 4 vols. (Philadelphia, PA: Lea and Blanchard, 1842), 1:302.
8 Lydia Goehr, *The Imaginary Museum of Musical Works: An Essay in the Philosophy of Music* (Oxford: Oxford University Press, 1992), 192.
9 William Weber, 'Did People Listen in the 18th Century?' *Early Music* 25, no. 4 (1997): 681.
10 Weber, 'Did People Listen,' 681.
11 Weber, 'Did People Listen,' 681.
12 Beresford, *Miseries*, 86.
13 Beresford, *Miseries*, 109.
14 C. Bruyn Andrews, ed., *The Torrington Diaries Containing the Tours Through England and Wales of the Hon. John Byng* (New York: Henry Holt, 1934–38), 4:133–34.
15 Adrian Tinniswood, *The Polite Tourist: Four Centuries of Country House Visiting* (Oxford: Basil Blackwell, 1989), 97; Ian Ousby, *The Englishman's England: Taste, Travel and The Rise of Tourism* (London: Pimlico, 2002), 50, 63.
16 Ernest Law, *The History of Hampton Court Palace, Vol. 3: Orange and Guelph Times* (London: George Bell and Sons, 1891), 327.
17 Law, *History of Hampton Court Palace*, 343.
18 William Fordyce Mavor, *The British Tourists; Or, Traveller's Pocket Companion*, 3rd ed., 6 vols. (London: R. Phillips, 1809), 2:235.
19 Quoted in Tinniswood, *Polite Tourist*, 104.
20 James Boswell, *Boswell's Life of Johnson*, eds. George Birkbeck Norman Hill and L.F. Powell, 6 vols. (Oxford: Clarendon Press, 1934–50), 3:161.
21 Tinniswood, *Polite Tourist*, 93. The various conditions contributing to the rise of polite tourism are strictly beyond the narrower scope of this chapter. Both Tinniswood and Peter Mandler, *The Fall and Rise of the Stately Home*

(New Haven, CT: Yale University Press, 1999), offer detailed, if divergent, accounts.

22 Stephen Clarke, '"Lord God! Jesus! What a House!": Describing and Visiting Strawberry Hill,' *Journal for Eighteenth-Century Studies* 33, no. 3 (2009): 365.

23 Quoted in Clarke, '"Lord God! Jesus! What a House!"' 365.

24 Walpole, *Letters*, 4:283.

25 Walpole, *Letters*, 4:283.

26 Walpole, *Letters*, 4:283.

27 Quoted in Constance Classen, 'Museum Manners: The Sensory Life of the Early Museum,' *Journal of Social History* 40, no. 4 (2007): 898.

28 Quoted in Classen, 'Museum Manners,' 898.

29 In 'Museum Manners: The Sensory life of the Early Museum,' Classen demonstrates how the problem of visitors touching exhibits is slowly disciplined through converting the museum into a largely visual experience.

30 Quoted in Rosalys Coope, 'The "Long Gallery": Its Origins, Development, Use and Decoration,' *Architectural History* 29 (1986): 59.

31 Quoted in Coope, 'The "Long Gallery",' 59.

32 Coope, 'The "Long Gallery",' 61.

33 Quoted in Coope, 'The "Long Gallery",' 62

34 See also Lawrence Stone, 'The Public and Private in Stately Homes of England, 1500–1990,' *Social Research* 58, no. 1 (1991): 229, who also considers the communicative aspect of the great house as a kind of speech act, 'a stunning public statement of power, wealth, and status.'

35 Quoted in Anne Goldgar, 'The British Museum and the Virtual Representation of Culture in the Eighteenth Century,' *Albion* 32, no. 2 (2000): 211.

36 As Adrian Tinniswood describes – or rather, reconstructs – the scene of polite tourism, visitors were content to 'note the picture which attracted him or her, offering little in the way of critical comment beyond the occasional "fine" or "very striking".' See Tinniswood, *Polite Tourist*, 110.

37 Helen Rees Leahy, *Museum Bodies: The Politics and Practices of Visiting and Viewing* (Farnham: Ashgate, 2012), 76.

38 Law, *History of Hampton Court Palace*, 326–27.

39 Thomas Roger Smith, *A Rudimentary Treatise on the Acoustics of Public Buildings* (London: John Weale, 1861), 37–38.

40 Smith, *Rudimentary Treatise*, 39.

41 Smith, *Rudimentary Treatise*, 50.

42 Smith, *Rudimentary Treatise*, 51.

43 Quoted in Smith, *Rudimentary Treatise*, 52.

44 Smith, *Rudimentary Treatise*, 52.

45 Smith, *Rudimentary Treatise*, 58–59.

46 Leahy, *Museum Bodies*, 28.

47 Leahy, *Museum Bodies*, 26.

48 Quoted in Leahy, *Museum Bodies*, 26.

49 William Hutton, *A Journey from Birmingham to London* (Birmingham: Pearson and Rollason, 1785), 189–91.

50 Hutton, *Journey*, 191–2.

51 Quoted in Leahy, *Museum Bodies*, 34.

52 Quoted in Leahy, *Museum Bodies*, 34.

53 Quoted in Leahy, *Museum Bodies*, 34.

54 Robin Evans, *Translations from Drawing to Building and Other Essays* (Cambridge, MA: MIT Press, 1997), 90.

55 Evans, *Translations from Drawing*, 77.

Bibliography

Primary sources

Andrews, Cyril Bruyn, ed. *The Torrington Diaries, Containing the Tours through England and Wales of the Hon. John Byng (Later Fifth Viscount Torrington) Between the Years 1781 and 1794.* 4 vols. New York: Henry Holt, 1934–38.

Beresford, James. *The Miseries of Human Life; or The Groans of Timothy Testy, and Samuel Sensitive.* London: W. Bulmer, 1806.

Boswell, James, *Boswell's Life of Johnson.* Edited by George Birkbeck Norman Hill and L.F. Powell. 6 vols. Oxford: Clarendon Press, 1934–50.

Hutton, William. *A Journey from Birmingham to London.* Birmingham: Pearson and Rollason, 1785.

Law, Ernest. *The History of Hampton Court Palace, Vol. 3: Orange and Guelph Times.* London: George Bell and Sons, 1891.

Mavor, William Fordyce. *The British Tourists; or, Traveller's Pocket Companion, through England, Wales, Scotland, and Ireland. Comprehending the Most Celebrated Tours in the British Islands.* 3rd ed. 6 vols. London, R. Phillips, 1809.

Smith, Thomas Roger. *A Rudimentary Treatise on the Acoustics of Public Buildings.* London: John Weale, 1861.

Walpole, Horace. *The Letters of Horace Walpole, Earl of Orford: Including Numerous Letters Now First Published from the Original Manuscripts.* Edited by John Wright. 4 vols. Philadelphia, PA: Lea and Blanchard, 1842.

Secondary sources

Clarke, Stephen. '"Lord God! Jesus! What a House!": Describing and Visiting Strawberry Hill.' *Journal for Eighteenth-Century Studies* 33, no. 3 (2010): 357–80.

Classen, Constance. 'Museum Manners: The Sensory Life of the Early Museum.' *Journal of Social History* 40, no. 4 (2007): 895–914.

Coope, Rosalys. 'The "Long Gallery": Its Origins, Development Use and Decoration.' *Architectural History* 29 (1986): 43–84.

Davidson, Jenny. *Hypocrisy and the Politics of Politeness: Manners and Morals from Locke to Austen.* Cambridge: Cambridge University Press, 2004.

Evans, Robin. *Translations from Drawing to Building and Other Essays.* Cambridge, MA: MIT Press, 1997.

Goehr, Lydia. *The Imaginary Museum of Musical Works: An Essay in the Philosophy of Music.* Oxford: Oxford University Press, 1994.

Goldgar, Anne. 'The British Museum and the Virtual Representation of Culture in the Eighteenth Century.' *Albion* 32, no. 2 (2000): 195–231.

Langford, Paul. 'The Uses of Eighteenth-Century Politeness.' *Transactions of the Royal Historical Society* 12 (2002): 311–31.

Mandler, Peter. *The Fall and Rise of the Stately Home.* New Haven, CT; London: Yale University Press, 1997.

Mee, Jon. *Conversable Worlds: Literature, Contention, and Community 1762 to 1830.* Oxford: Oxford University Press, 2011.

Ousby, Ian. *The Englishman's England: Taste, Travel and the Rise of Tourism.* London: Pimlico, 2002.

Stone, Lawrence. 'The Public and the Private in the Stately Homes of England, 1500–1990.' *Social Research* 58, no. 1 (1991): 227–51.

Tinniswood, Adrian. *The Polite Tourist: Four Centuries of Country House Visiting.* Oxford: Basil Blackwell, 1989.

Weber, William. 'Did People Listen in the 18th Century?' *Early Music* 25, no. 4 (1997): 678–91.

Part III

Sound, noise and the incivility of the crowd

6 The sound of the spirit

Auditory enthusiasm and the attack on Methodism in the eighteenth century

Peter Denney

'The *Importance* of *Sound* and *Noise* in the *Gospel* of *Methodism* is truly great,' proclaimed an anonymous author, probably William Combe, in a caustic satire on this popular religious movement. Published in 1778, Combe's satire everywhere implied that Methodist spirituality had an acoustic orientation. From loud extempore preaching and voluble prolix prayer to inarticulate vocalisations like shouting and crying, Methodist worship was portrayed as a thoroughly noisy affair. Moreover, the alleged sonic disorder of Methodism made it the epitome of religious enthusiasm, threatening civility, reason and social stability. According to Combe, the noisy quality of this form of enthusiasm was linked to the plebeian status of its adherents, since the labouring classes found in evangelical acoustic practices an outlet for their fondness for raucous behaviour. Upon hearing the 'joyful News' of salvation, for example, these vulgar Methodists opted for inspiration over industry, and, leaving their workshops and mines, started singing 'Hallelujahs' instead of 'Ballads.' In addition, they regarded learning as an impediment to zeal, for their worship emphasised strong emotions which, free from 'Reason,' tended to 'give their *Frenzies* wing, | *Groan, weep, rave, rant, confess, exhort*, and *sing*.' Due to this conduct, Combe concluded that, among Methodists, noise was a sign of godliness, with praise being reserved for those with 'Tongues that *rave* and *flow*.' As a result, prayer was a voluble activity, a kind of 'rapt'rous Talk' with an audible, omnipresent God. Similarly, preaching was a noisy event, as popular orators treated spoken words as missiles to be fired at their auditors. By discharging their 'Thunder,' enthusiastic preachers aimed to induce an 'assenting Groan' in the 'Mob' or provoke some other impulsive, acoustic response. Alarmingly, in Combe's view, such 'oratorical Graces' made Methodists resemble the 'Tub-Orators' of the Civil War, because the saints of the commonwealth also used groaning, crying and other vocalisations to incite the passions of their listeners. For Combe, 'Scripture' constituted the 'Rule of *Manners*,' and the noisy style of Methodist devotion functioned to 'vindicate *Indecency*' by misinterpreting the holy book. Referring to a 'fiery' Methodist preacher, who never entered 'a Pulpit without lifting up his *Voice* like a *Trumpet*,' for instance, he noted that this 'boist'rous' style of speech, perhaps based on a belief in the power of sound, as conveyed by the story of Joshua at Jericho, contradicted an important biblical principle. For unlike Methodism, the bible taught that 'there came a *great Noise* and a *Whirlwind*, but the Lord was *not there*; and afterwards there was a *still small Voice*, and the Lord *was there*.'[1]

The 'sensory turn' in historical scholarship, as superbly documented by Mark M. Smith,[2] has generated several excellent studies of the acoustic aspects of evangelical religion. These studies have focussed primarily on nineteenth-century America, examining only in passing the different context of eighteenth-century Britain. Nevertheless, they have all noted the sonic orientation of Methodism, and the role of sound in negotiating the boundary between enthusiasm and legitimate religious feeling. In a remarkable work, Leigh Eric Schmidt has highlighted the 'auditory intensity' of popular evangelical piety, as inner voices combined with external vocalisations to make devotion a noisy affair in which sounds facilitated an interpenetration of earthly and sacred space.[3] But due to the authority of rational religion, the sounds of revival were heard by the educated elite as signs of mental or emotional derangement. As a result, enthusiasm came to be regarded as a form of noise pollution, which emanated from an unruly, brutish, irrational crowd. In reaction to this, as Anne Taves has shown, evangelical leaders on both sides of the Atlantic aimed to define true religion against enthusiasm, though John Wesley, among others, retained a belief that some involuntary sounds, even loud incoherent utterances, were genuine responses to the witness of the Spirit.[4] Plainly, while there were many different ways of listening and responding to God, intense auditory activity became a defining feature of popular religious zeal. Thus, in an elegant essay on the sensory aspects of religion in eighteenth-century Britain, Phyllis Mack has underlined the noisy character of Methodism and evangelical Dissent, relating this emphasis on sound to the attempt to enliven piety through expressions of strong emotion.[5]

These studies have enriched our understanding of Methodism and evangelical religion, illustrating the benefits of an acoustic approach to such topics. In addition to highlighting the importance of sound as a bearer of social and spiritual meaning, they have demonstrated that elite antipathy to the noise of popular religious feeling was related to the fusion of theology and empiricism, which flourished from the late seventeenth century. Increasingly, religious experiences were given natural explanations, with prophetic words, involuntary vocalisations and other aural expressions of devotion linked to an agitation of animal spirits or vibrating nerves. And yet the attack on the noise of Methodism was also shaped by the ideology of politeness, as it emerged in Britain in the eighteenth century.

This essay will examine representations of noise in Methodist and anti-Methodist literature, especially in relation to ideas of enthusiasm and civility. It will argue that the association of raucous piety with subversive enthusiasm was informed by a dominant version of religion, which conflated politeness and Christian morality. But defenders of Methodism, including John Wesley, had an ambivalent attitude to noise. Seeking to promote religious zeal, they heard cries, shouts and allied exclamations as expressions of possible spiritual vitality as well as potential social disorder, authenticity as well as incivility.

In the eighteenth century, of course, most representations of Methodism as noisy were hostile to this religious movement. Beginning in the late 1730s as a revival movement within the established church, it was heard as an assault on the dignity of religion, not least because it produced a new set of acoustic values and practices. This challenge to the conventional sonic aspects of Anglicanism was partly a result of the fact that Methodism appealed to the labouring classes. Specifically, as plebeian voices were raised in song, conversation or spiritual experience, they were perceived to undermine the solemn and rational nature of piety, lowering its tone by increasing its volume. For anti-Methodist writers, as we shall see, the vehement voices and disorganised sounds of evangelical worship were heard as evidence of a dangerous enthusiasm. In fact, in the opinion of Theophilus Evans, the noisy character of popular evangelicalism constituted a 'revival' of the violent sectarianism of the Civil War of the mid-seventeenth century.[6]

Since a major impetus to the spread of politeness in the eighteenth century was the attempt to prevent a repeat of the social disruption caused by religion in the recent past, it was not surprising that there emerged a form of refined piety hostile to the presence of noise in Christian worship. Around this time, the Anglican Church was fashioned into a distinctively polite institution,[7] partly by removing or regulating a wide variety of sounds. Non-verbal vocalisations were disavowed, while both the tone and content of spoken words were scrutinised. Conversely, noisy outbursts in religious settings were derided for smacking of enthusiasm. This process commenced during the Restoration with the development of latitudinarianism, which, with its insistence on sobriety, moderation and practical morality, generated a style of mainstream Anglicanism as averse to enthusiasm as it was conducive to politeness.[8] There was no place in this religious tradition for vehemence, verbosity or other noisy expressions of devotion. For these qualities implied a self governed by extreme or wayward passions, leading to volatility and contention. As one latitudinarian minister noted in the 1670s, Christianity did not involve 'letting our Tongues loose, but in bridling both them and our exorbitant Affections.' The aim of such refined worship, he concluded, was to 'make less Noise, be less Disputatious, and more Obedient; that we Talk and Cavil less, and Be and Live better.'[9] As religion was disciplined by politeness, transformed into an orderly, conciliatory force, propriety came to be a demonstration of piety, attenuating the acoustic component of devotion.

Once disciplined, religious worship was represented as enabling individuals to fashion a polite social identity,[10] not least because it fostered a habit of reflection incompatible with noisy conduct. Paraphrasing Richard Steele's famous definition of politeness, a contributor to *The Spectator* remarked that the 'Business of Religion' was 'not so much to extinguish our Passions, as to regulate and direct them to valuable well-chosen Objects.'[11] Accordingly, polite and religious practices became intertwined, sharing the same purpose of restraining the passions and refining them into manners. For Steele, meekness was the 'Characteristick of a Christian' no less than a connoisseur, and, intriguingly, this virtue was associated with a fondness for 'silent Awe.'[12] One way of making Christianity conducive to refinement was by representing devotion as a quiet, contemplative activity. As Joseph Addison noted, the 'Contemplation of the Divine Being' involved a frame of mind averse to 'dissolute Mirth' and 'Levity of Behaviour,' replacing such qualities with 'perpetual Serenity' and a 'habitual Inclination to please others.'[13] In contrast to the divine immediacy of enthusiasm, the polite religion advocated by *The Spectator* encouraged individuals to take a distanced view of God by contemplating his Creation, perceived as a work of art. To contemplate the Creation, wrote Addison, was to recognise the 'ineffable Goodness' of its Author, a thought that he believed was 'too big for the Mind of Man, and rather to be entertained in the Secrecy of Devotion and in the Silence of the Soul, than to be expressed by Words.'[14] As envisaged by the proponents of politeness, devotion was most perfectly expressed by speechlessness, by being lost for words in response to the grandeur of God. While there was value in religious warmth, this was only positive if it appeared as quiet reverence rather than noisy activity. Shouting, crying and haranguing were violations of taste no less than piety.

Due to this conflation of religion and politeness, the noise of Methodism was heard not only as a sign of the vulgarity of its plebeian followers,[15] but as evidence of their menacing enthusiasm. In 1739, for example, the clergyman and literary critic, Joseph Trapp, depicted the revival of religious zeal as an affront to polite piety. With Calvinist Methodists in mind, Trapp focussed on the '*outrageous* Anger' of such enthusiasts, whose tendency to turn 'just Indignation into rude Invective' was 'inconsistent with the Virtue of Meekness.' To counter this 'false' zeal, he proposed that religious feeling be disciplined by avoiding 'intemperate Passion' and excitable speech so as to exclude 'any Thing *contrary* to *Decency*, and *good Manners*.'[16] A degree of emotion in religion was positive, he noted, but, when taken to excess, this caused rage and impetuosity, antisocial qualities, which posed an acute problem in the labouring classes. 'The good Christian,' concluded Trapp, 'may have … some Warmth, and even Heat; but not be like a Red-hot Iron, hissing, and sparkling from the Forge, and dropping Fire wherever it reaches.'[17] This was a revealing metaphor, for, by comparing enthusiasm to a 'hissing' iron, the combustible, unintelligible noise of plebeian Methodists was linked to their status as mechanical workers with uneducated minds.

For proponents of politeness, the most troubling sounds were the cries, groans, shouts and other vocalisations, which emanated from public assemblies. Along with convulsions, these loud, involuntary sounds were central to the spirituality of Methodists.[18] For believers, they were responses to an indwelling Holy Spirit, signs of a direct encounter with an interventionist God. Such vocalisations were especially associated with conversion, a bedrock of Methodist religiosity.[19] Conversion gave enormous energy to revival meetings, as participants could hear the power of God in their midst. Provoking intense feelings and vigorous sounds, it involved a dramatic conviction of sin, followed by an exhilarating assurance of rebirth, in a process marked by noisy exclamations of anguish and, then, elation. As Misty Anderson has argued, critics of Methodism perceived this as a threat to civilised subjectivity, because it posited that personal identity could be realised by self-abandonment rather than self-control, eschewing rational thought and polite conduct,[20] while licensing extreme acoustic and emotional responses, from crying to shouting. 'Quit yourselves like Men,' urged Alexander Jephson, 'and be not hurried away with Noise and Clamour, but exercise that Reason and Understanding with which God hath endued you.' Despite his qualified approval of religious warmth, Jephson was adamant that the 'dreadful Shriekings and Groanings' of believers were 'deservedly Objects of Scorn,' grounded as they were in the delusions of vulgar preachers.[21] Far more hostile was the author of *Fanatical Conversion*, who depicted Methodists as victims of madness, their unregulated passions causing diverse forms of noisy behaviour:

> Laugh, cry, display what Modesty should hide;
> Now boil with Rage; *becalm'd*, again, subside; …
> These *Mad-folks* foam, rank, caper, and curvet,
> Flame, shiver, tremble, dance, chaunt, rave, and fret.[22]

This disavowal of noisy worship was reinforced by the development of a medicalised theory of enthusiasm. Like a number of critics, George Lavington interpreted the 'wild Noise' of Methodist devotion as a symptom of hysteria, an affliction which caused sufferers to endure 'alternate Fits of Joy and Sorrow, Laughing and Crying.'[23] While hysteria, hypochondria and melancholy had been transformed into fashionable medical conditions by the mid-eighteenth century,[24] indications of the sensibility of polite individuals, they were thought to be perilous in the labouring classes who were considered to externalise rather than internalise excessive emotions, causing the wild, noisy behaviour of the crowd.[25] Other critics heard something heretical as well as pathological in the vocalisations of Methodists, and it was agreed that the movement eroded the civilising, conciliatory aspect of religion. Such utterances, noted Thomas Mortimer, replaced a view of the Holy Spirit as peaceable with a notion of 'divine inspiration' based on 'heresies, cursings, and revilings.' Of this 'modern spirit,' he added 'it alternately sings, it groans, it sighs, it weeps, it thunders out damnation, it bawls in tabernacles, it raves in our streets, it exclaims in our houses.'[26] Without exception, opponents of Methodism dismissed these expressions of religious zeal as sounds without meanings, mixtures of noise and nonsense, insanity and ecstasy.[27] Furthermore, the links between noise, enthusiasm and irrationality enabled critics to claim that Methodist religiosity was sensual rather than intellectual, promising to reduce humankind to the level of the brute creation. Accordingly, several writers likened Methodists to animals, their 'hideous Noise' resembling, for instance, the 'Grunts' of a 'Hog' or the whinings of a 'midnight Cat.'[28] In comparable fashion, their vocalisations were loaded with crude bodily associations, such as when James Adair compared groaning to farting.[29] Whether Methodism was interpreted as heresy or pathology, mass emotionalism or gross sensualism, its extreme, involuntary sounds were heard as a failure of auditory communication, which, by threatening the existence of refined religion, challenged, too, the order of society and the coherence of the self.

Alongside unintelligible sounds, popular evangelical piety was also marked by an openness to divine voices. The minds of followers were compared to echo chambers, reverberating words from the Holy Spirit, whether a quotation from scripture or a reply to prayer.[30] For most anti-Methodist writers, of course, this constituted a form of sensory derangement, as a surge of animal spirits interfered with the processes of perception, preventing reflection from curbing the imagination. As Lavington noted, such 'Voices' were the 'Echo' of a 'distempered Fancy.'[31] 'Irregular and turbulent Commotions of the Blood,' he explained, 'drive out the Reason and Senses; the Persons thus disordered are filled with a thousand Chimaera's; fancy they *hear* and *see* and *feel* Things, which have *no Existence in Nature*.'[32] Worryingly, this susceptibility to receive messages from God was believed to be incompatible with meekness, moderation or mutuality. According to William Penrice, for example, the 'divine communications' of Methodism supported 'monstrous notions,' which were conveyed using 'extravagant words and gestures.'[33] From this perspective, hearing divine voices was regarded as part of an acoustic feedback loop, for such audible spiritual impressions were often occasioned by vehement preaching, just as they frequently generated noisy expressions of worship. Little wonder that John Mainwaring derided the tendency of 'modern Mystics' to mistake the written words of scripture for the spoken words of God. For this was proof that Methodist piety appeared to 'draw all its strength and nourishment from *Sounds*.'[34]

Hymn-singing was another feature of Methodist worship which critics attacked for promoting noise, emotional excess and self-abandonment.[35] Probably no religious practice was more important to the development of Methodism than singing. As David Hempton has argued, the hymns composed by Charles Wesley and, to a lesser extent, by his brother John were fundamental to evangelical spiritual experience.[36] In addition to transmitting core religious beliefs, these hymns linked devotion to the expression of strong, personal feelings.[37] Accordingly, singing enacted and reinforced an emotional style of piety. The centrality of hymn-singing was also a source of the substantial appeal of Methodism in plebeian circles, not least because the Wesleys set religious lyrics to popular tunes.[38] As William Vincent remarked in a discussion of Methodist Psalmody, 'it is not rashness to assert, that for one who has been drawn away from the Established Church by Preaching, ten have been induced by Music.'[39]

The role of singing in inciting those passions which caused religious warmth to erupt into enthusiasm was a major theme in anti-Methodist writing. In another acoustic feedback loop, critics claimed that Methodist hymn-singing stimulated violent passions, while these passions then incited vocalisations and convulsions. Describing enthusiasm as 'Holy Madness,' John Douglas offered an evocative first-hand account of this sonic mayhem. 'I myself, have seen the singing of Psalms have so violent an Effect on an enthusiastic Devotee,' he noted,

> that, being heated to a Frenzy, she trembled with Agony, in a Manner foamed at the Mouth, and uttered such hideous Cries that it became necessary to gag her, and carry her out of the public Assembly where this happened.[40]

Lavington also believed that Methodist singing was a stimulant to errant enthusiasm, whether madness or possession. Revealingly, too, he regarded Methodist music as equivalent to noise. 'If something *Extraordinary* was not expected from *Noise*,' he asked, 'what Occasion of being *so very loud?*'[41] For Lavington, it was the loudness of Methodist hymn-singing, which made it so dangerous, kindling the passions with such intensity that they eluded the power of reason to regulate them.

Enthusiasm was held to be a contagious phenomenon,[42] and numerous critics of Methodism believed that hymn-singing contributed to the spread of this psychological and social disorder. Such contagion was linked to the power of sound to rouse the passions and, once agitated, transmit them between individuals within a certain acoustic environment. An important examination of this process was undertaken by the Anglican vicar Thomas Green in an anti-Methodist tract, published in 1755. Green drew on the affective theory of music formulated by Charles Avison to argue that there was a close connection between sound and the passions.[43] This was a valuable aid to religious worship if, via moderate singing or preaching, it produced a form of 'devotion ... both *warm* and *rational*.' But when sound became noise, when it became too loud or too irregular, it bypassed reason and generated 'sudden emotions or transports' which, in a religious context, could be mistaken for the 'workings of the Spirit.'[44] Because sound had a direct impact on the nerves, it also triggered bodily convulsions. Furthermore, the 'infectious' nature of enthusiasm had a vital acoustic component. As Green averred, 'hearing the noise' of convulsions was one way of contracting them.[45] This theory regarding the role of sound in transmitting enthusiasm was based on a pamphlet by the Dutch minister Johannes Stinstra, translated into English in 1753. Stinstra compared enthusiasm to two well-tuned violins: when the string of one instrument was touched, the corresponding string on the other instrument would also resound. In the culture of sensibility, this was a common symbol of sympathetic exchange.[46] Indeed, the metaphor of resonating violins was grounded in an awareness that sound constituted an important medium of emotional contagion. In the context of collective religious feeling, tones of voice and styles of music were both regarded as crucial vehicles for spreading enthusiasm, and, even against their will, individuals could be influenced by sound, especially noise, to share the passions of those around them. With this in mind, Stinstra argued that the 'Spiritual Hymns of the *Fanaticks*, sung in an agreeable Manner in moving Airs, are one of the best Means to propagate *Fanaticism*.'[47]

The labouring classes were considered to be exceptionally vulnerable to enthusiasm because their proneness to extreme passions made them unusually attracted to music, noise and other acoustic phenomena. It was a 'proverbial Saying,' remarked John Scott, that 'The Mob will follow the Fiddle.'[48] In the critique of Methodist hymn-singing, it was argued that popular tunes interested the labouring classes in religion by courting their enthusiasm. 'People of weak Minds' were fond of this 'frothy Way of Singing,' complained William Riley, but it diminished rather than strengthened devotion, introducing a 'Variety of ludicrous Sounds,' which 'excited' 'silly and absurd' passions.[49] Whereas traditional church music stilled the mind, it was claimed, congregational singing fostered a noisy religious culture, which was incompatible with reason, sobriety and self-control.

Methodist preaching, too, was criticised as a noisy activity in which the speaker used a loud, impassioned voice to incite intense audible responses in listeners, forging a disorderly sonic community. To quote Theophilus Evans, 'the boisterous Manner of Delivery so much affected by the Methodist-preachers … may with great Justness and Propriety be adjudged the principal Cause of all these Disorders, viz. their Roaring, Yelling, bodily Concussions and Agonies.' Enthusiasm was regarded by Evans as an acoustic phenomenon in which sound, often beginning with fervent preaching, enabled the inflammation and exchange of passions in the crowd. And to emphasise the point, he highlighted the infectious nature of such public expressions of noise, noting that 'when *one* falls into *Fits*, or *screams*, it affects the rest by Sympathy, or quickly *one* after *another*, as it were by Contagion.'[50] Like singing, Methodist preaching was heard as an example of the power of sound, not least the ability of noise to kindle intense, unruly collective emotions.

The attack on noisy Methodist preaching often centred on John Wesley and, especially, George Whitefield. Among both supporters and opponents of Methodism, Whitefield was well known for his loud, penetrating voice, and Benjamin Franklin estimated that he could be heard by an outdoor crowd of 30,000 people.[51] Combining emotionalism with fiery Calvinism, he perfected a dramatic preaching style which alternated between pathos and hellfire denunciation, while deploying plain language, extemporaneity and bodily gesticulation to create a striking intimacy of tone.[52] According to Cornelius Winter, Whitefield's sermons usually included the 'unreserved use of tears,' and he exhibited his strong feelings by stamping his feet 'loudly' during bouts of weeping.[53] This electrifying style of pulpit oratory was criticised by anti-Methodist writers, who associated its noisiness with insincerity, indecency and rabble-rousing. For Samuel Johnson, for example, the charismatic preacher had a 'mind like a hammer,' and, as a result, he 'vociferated' rather than talked, conveyed impressions instead of thoughts and practised a style of 'oratory' suitable only for the mob.[54] Similarly, in 1769, Thomas Chatterton described Whitefield's preaching as 'Noise and Nonsense,' a mixture of 'Rowling, Growling, Rumbling and Grumbling,' in which 'cries' of damnation reverberated 'around,' making the 'long-eared rabble grunt in dismal sound.'[55]

Another reason for the representation of Whitefield as a scourge of politeness was his initiation of itinerant field preaching, a practice perceived to undermine acoustic spaces associated with social stability. If Methodist leaders continued to 'ramble up and down singing Psalms, and preaching in the open Streets, or in the more open Fields,' warned one commentator in 1739, 'wanton Curiosity will carry Thousands to see and hear such new Things,' and 'hundreds of the ignorant Multitude will innocently be corrupted.'[56] Field preaching generated a noise from which no one within earshot could escape. And this sonic disturbance was believed to attract as well as antagonise individuals, disrupting the performance of their duties. 'To pray, preach, and sing Psalms, in the Streets and Fields,' complained Joseph Trapp, reduced the Christian religion to 'Riot, Tumult, and Confusion.'[57] From this perspective, the creation of outdoor evangelical soundscapes, where loud, passionate preachers stimulated noisy, spiritual experiences in believers, undermined the basis of both Anglican piety and polite sociability. Contributing to this supposed cacophony was the low social class of itinerant field preachers, whose lack of education, it was argued, made them mistake emotion for inspiration and vehement speech for godly, heartfelt oratory. Far from being an expression of religious ardour, wrote Thomas Green, a 'loud or uncommon vehemency of voice' was opposed to the 'mild composed voice' required by the 'meek spirit' of Christianity.[58] As a result, the noise of popular preachers became a symbol of their vulgar status, their alleged inability to regulate their passions and refine them into manners.

When critics attacked the noise of Methodist preaching, then, they were referring primarily to its loudness, though this was attributed to a range of qualities, from verbosity to vehemence, melodrama to meaninglessness. Of the many misdemeanours of Methodist preachers, one supporter of established religion singled out their 'vehement and bawling vociferations,' their 'dismal and melancholy sighs and groans' and their 'bombastic and affected sounds and gestures.'[59] As we have seen, the danger of such noisy oratory lay in its capacity to inflame the passions of auditors, prompting spasmodic acoustic and bodily responses. The result was the transformation of listeners into an unruly mob, able to be manipulated or incited by an indecorous preacher using potent, theatrical sounds.[60] As satirised by Nathaniel Lancaster, the booming voice of a typical hellfire preacher functioned to whip listeners into a state of frenzy:

> So loud he thunders – deals about his flames –
> With such terrific force he strikes the ear –
> So shocks the senses – so confounds the soul –
> That frighted Reason her dominion quits;
> And Frenzy fills the abdicated throne.[61]

Women and the labouring classes were deemed to be particularly vulnerable to noisy oratory, because, despite differences in gender and status, both groups were held to respond to rhetoric emotionally rather than intellectually. And as the volume of the speaking voice increased, so reactions to it were thought to become more emotionally turbulent. On this point, Green cited the model of polite eloquence espoused by John Mason, who warned of the 'disagreeable' effects of listening to a speech in which the voice was 'too loud.' Since such communication did not appeal to the understanding, the attempt to 'perswade or affect by mere Vehemence of Voice' was denounced as a form of 'false Oratory.' In fact, Mason argued that 'none but the most low, weak, and mechanical Minds will be affected with mere Dint of Sound and Noise.'[62] At the same time, around the middle of the eighteenth century, proponents of polite religion began to assign an important place to the emotions in pulpit oratory. Fuelled by the culture of sensibility, this stress on expressive oratory was deployed to rejuvenate Christian worship.[63] And yet religious writings insisted that preaching conform to a standard of polite taste, which moved the calm affections of auditors without mobilising any unruly passions. This was 'true eloquence,' declared David Fordyce, whereas to raise violent, instantaneous impulses of joy or sorrow, anger or compassion, was 'puerile Oratory,' since it could be achieved by 'Sound as readily as by Sense.'[64] In the context of preaching, then, noise indicated the crossing of a threshold between desirable religious warmth and illegitimate enthusiasm. As an accompaniment to popular evangelical devotion, it was also regarded as a means of depriving listeners of their autonomy, inciting turbulent passions which dissolved the self into a manipulable, if unpredictable, crowd. With the 'most powerful vociferation,' fretted Thomas Mortimer, the 'enthusiastic preacher ... works on the passions of the gaping multitude, till the whole congregation resounds with sobs and sighs; and spiritual groanings.'[65] Predictably, such preaching was classified as 'low and unmanly,' and the disorderly, acoustic responses it generated were depicted as feminine or plebeian failings, the result of extreme passions as well as the inability to regulate them.

Among adherents of Methodism, however, the noise of evangelical piety was rarely defended in unequivocal terms, and often it was utterly rejected. In fact, Methodists frequently insisted on the civility of their style of devotion, emphasising its quiet character. As Brett McInelly has shown, the views of both leaders and followers of the movement were shaped by anti-Methodist writings.[66] Following John Wesley, they refuted accusations of enthusiasm, and this sometimes entailed defending their principles and practices in the language of politeness. As we saw earlier, Wesley was repeatedly attacked for encouraging intense acoustic experiences among followers. And partly in response to such reproaches, he developed an ambivalent attitude to the role of noise in religious worship.

Despite a preference for exciting over indifferent worship, Wesley believed that discipline had to temper zeal, instilling a love of quietness in adherents.[67] Nevertheless, his opponents overlooked this commitment to sonic moderation and emotional equilibrium. In fact, he was berated for equating noise with spiritual vitality. With typical disdain, one critic condemned him for perceiving a 'decent, attentive, *quiet* Congregation' as lifeless rather than courteous.[68] Likewise, Bishop Gibson censured both Wesley and Whitefield for rejecting 'serious and composed' worship and interpreting as evidence of the Holy Spirit those 'sudden Agonies, Roarings and Screamings, Tremblings, Droppings-down, Ravings and Madnesses,' which affected their auditors.[69] For Wesley, religious zeal was an essential component of devotion. But he recognised that a spurious version of this 'fervent love' had caused mischief in the past by licensing strong, wayward passions such as animosity.[70] In consequence, he insisted on the close relationship between zeal and meekness, which, when combined, made quietness an important feature of Christian worship. Specifically, meekness subjected zeal to self-control, minimising noisy activity and fostering 'calm, dispassionate' behaviour.[71] In this way, Wesley followed Steele in regarding meekness as a quiet quality which was conducive to politeness no less than religion.

This representation of Methodist religiosity as compatible with refined conduct gave rise to a repudiation of many different sorts of noise. In 1774, for example, Wesley asked Hannah Ball to warn their friend, the preacher Joseph Bradford, 'not to speak too fast or too loud.'[72] And when Wesley himself preached to the Methodist society at Norwich in 1761, he disapprovingly noted that the large congregation created a 'hubbub,' just like an unruly mob. Revealingly, however, the auditors stopped this racket and listened 'quietly and silently' once they had been reprimanded by their leader.[73] Indeed, Wesley's ability to transform a noisy crowd into a quiet, orderly congregation was a common trope in his journal, and he associated this change with the spread of true devotion, restrained without being unmoved. An excellent instance of this was his reformation of the Methodist meeting in Beech Lane, London, where, before his arrival, a mixture of profanity and enthusiasm made religious practices resemble noisy popular sports. But Wesley rectified this acoustic licentiousness by relocating the meeting to the principal Methodist chapel, the Foundry, so that it could be placed under strict surveillance:

> 1. The meeting in Beech-lane, before I came to town, was like a bear-garden; full of noise, brawling, cursing, swearing, blasphemy, and confusion. 2. Those who prayed were partly the occasion of this, by their horrid screaming, and unscriptural, enthusiastic expressions. 3. Being determined either 'to mend them or end them,' I removed the meeting to the Foundery. 4. Immediately the noise, brawling, cursing, wearing, blasphemy, and confusion ceased. 5. There was less and less screaming, and less unscriptural and enthusiastic language.[74]

While opponents implied that the noise, excess and disorder of Methodist religious practices made them indistinguishable from carnivalesque festive rituals, Wesley argued that evangelical devotion had a civilising effect on local communities, replacing idle, sensual pastimes with edifying spiritual pursuits. This disavowal of customary habits of conduct was one reason why the indomitable preacher was often subjected to the noisy protest of the crowd.[75] Everywhere he went, Wesley was met by groups of plebeian protesters, who used sound as a weapon to defend their customary culture. Often supported by members of the patrician elite, these agitators sought to silence Methodism by using established sonic rituals of protest from shouting to discordant music-making. Thus, bell-ringing became a strategy for drowning out Wesley's voice while he preached.[76] Likewise, shouting and cursing were standard forms of obstruction. At Epworth in 1763, a local gentleman paid a mob to shout out insults, bawl out obscenities and play the French horn during Wesley's oration, though the congregation here, apparently, remained calm and 'quietly attended to the end.'[77] Throughout his journal, revivalism was linked to an improvement in morality, which had a distinctively acoustic register as noise was replaced by quietness. When Wesley visited Barrow upon Humber, for example, he recalled that, on a previous visit, the 'mob was exceeding rude and noisy.' Due to the spread of Methodist religiosity, however, a moral reformation had occurred in the village, so that 'all the people were now quiet and attentive.'[78] This moral reformation could even happen instantaneously. On numerous occasions, Wesley described the way in which a raucous crowd, or individual, disrupted his preaching, only to be rendered quiet, almost speechless, by the end of his sermon. At York in 1752, for instance, a hostile crowd had been incited to throw 'bitter curses' at Wesley and his followers, but from the time he started preaching, the chapel became filled with the 'presence of God' and the 'opposers opened not their mouths.'[79] A fascinating variation of this process took place during his sermon at Beverly in 1784, when a town crier kept shouting until his wife gagged him 'so that he could not get out one word.'[80] But while it was not unusual for followers to silence protesters in this way, Wesley mostly attributed any sudden adoption of quiet behaviour to the power of God, working through his message of repentance, rebirth and Christian perfection.

According to Wesley, however, Methodism did not so much abolish sound as regulate it, leading to a more orderly, less noisy culture. Specifically, evangelical piety replaced a licentious, pagan soundscape with a disciplined, Christian soundscape, as believers came to value self-reflection over self-abandonment. At Christmas in 1742, for example, Wesley preached on Psalm 81 in order to show that the phrase, 'make a cheerful noise unto the God of Jacob,' did not justify rowdy public revelry. On the contrary, the proper way of observing a religious holiday involved public and private prayer, talking about the majesty of God, and 'hearing, reading and meditating on his word.'[81] For Wesley, evangelical devotion included sound, but reflection had to regulate speech, curtail behaviour and thus contribute to an atmosphere of quietness.

Wherever he preached, then, Wesley heard a quiet audience as evidence of the success of his Methodist message. Above all, he associated quietness with attentiveness, proof that his followers were able to eschew the distraction which was a hallmark of enthusiasm. Judging by the myriad times that Wesley described a congregation as 'quiet and attentive,' this was some of the highest praise he could bestow on a religious gathering. Preaching at Thirsk in 1784, for instance, such an assembly provided confirmation that God had 'touched many hearts.'[82] Similarly, when he addressed a 'quiet and attentive' congregation at Hertford in 1763, he became confident that the community would soon experience a spiritual revival and the associated moral reformation.[83] Of course, quietness was also important for the obvious reason that, as Wesley discovered, it was a precondition for hearing his voice, especially when the congregation was large. In addition, only in quietness, he wrote, could believers hear the voice of God, 'who loudly ... calls all his children to love one another.'[84] While opponents of Methodism regarded this apprehension of God's voice as a case of enthusiasm, Wesley was able to interpret it as compatible with civility since it turned devotion into a quiet affair, with listeners embracing silence in order to hear divine words.

In hostile territory, however, Wesley never shirked away from entering into spiritual battle with a noisy, riotous assembly, and his preference for quiet piety did not dampen his zeal for the salvation of souls. 'I love quietness and silence as well as ever,' he declared, 'but if I am called into noise and tumult all is well.'[85] His critics, of course, claimed this as proof that he possessed a penchant for disorder and noisy activity. But he evidently perceived any engagement with a hostile crowd as a sacrifice of his own quiet spirituality, made with some reluctance for the sake of increasing order as well as devotion in the populace.

As a proponent of the witness of the Spirit, however, Wesley refused to explain religious experiences in psychological terms, not least because such noisy, bodily responses to God could be justified by biblical precedents.[86] This was a point he made in *A Farther Appeal to Men of Reason and Religion*, written in 1746. Specifically, Wesley argued that repentance and redemption required the cooperation of the Holy Spirit, and this could be accompanied by noisy expressions of anguish and joy, remorse and release, which thus had to be tolerated or welcomed. Although the 'Remission of Sins' might not always be accompanied by 'Sudden Agonies, Roarings, Screamings, Tremblings, or Droppings down,' declared Wesley, 'if God is pleas'd at any Time to permit any of these, I cannot hinder it.'[87] He elaborated on the point more fully in his journal in November 1759 after preaching at Everton, Bedfordshire, a place he had visited earlier in the same year. During his first visit, in May, the congregation had responded to his message with trances, cries and convulsions, but now, six months later, only some people 'trembled exceedingly,' and the volume of the meeting had dwindled to a 'low murmur.' This led Wesley to surmise that the 'danger' was not to 'regard extraordinary circumstances too much, such as outcries, convulsions, visions, trances,' but to 'regard them too little; to condemn them altogether; to imagine they had nothing of God in them, and were a hinderance to his work.'[88] Aural expressions of spiritual intensity were evidently desirable, if not essential, components of worship. Wesley acknowledged that these acoustic and bodily manifestations of religious zeal might be perverted or fabricated, but the prospect of occasional error was a cost he was willing to pay to preserve belief in the power of the Holy Spirit, especially its role in facilitating atonement.[89] Certainly, he remained attached to quietness as a stimulus to both reverence and civil order. But this preference was in tension with his commitment to a noisy evangelical soundscape as animated as it was unrefined.

In fact, a quiet congregation could be heard as torpid rather than tranquil, devoid of religious warmth rather than overflowing with polite reverence. At a small village near Milford Haven, for instance, Wesley noted that the 'people, one and all, seemed as dead as stones, perfectly quiet, and perfectly unconcerned.'[90] But his preaching soon stirred their hearts, and the sound of bitter weeping registered the presence of God and the genuine repentance of new believers. Nothing could save the genteel folk at Stow-on-the-Wold, however, for they constituted a 'very dull, quiet congregation,' unresponsive to the intense spirituality of Methodism.[91] As demonstrated by this reference to social class, Wesley associated quietness with elite status. And yet the quiet conduct of such a refined congregation signified ignorance and insipidity. It was as if the quietness of gentility had the potential to muffle rather than moderate religious feeling. In the marketplace at Frome, for example, Wesley preached to a congregation comprising mostly gentlemen, who exhibited their polite status by maintaining a quiet, disinterested disposition during worship. But rather than praise their attentiveness, he described them as 'inferior' to the exuberant keelmen and colliers of Bristol, passionate about their Christian faith, though not polished in their articulation of it.[92] The famous evangelical preacher still regarded quiet worship as, primarily, a positive activity, a reconciliation of decency and devotion, but, at the same time, he was aware that composure could mutate into coldness. From this perspective, a quiet congregation could be heard as an example of incivility no less than courtesy, aloofness no less than reverence.

Like Wesley, the adherents of Methodism perceived sound to be an important aspect of evangelical piety, though they regarded noisy worship with some ambivalence. Certainly, many believers followed their leader in seeking to confer respectability on their religious zeal by depicting it as quiet.[93] Nevertheless, the noise of devotion was also represented as conducive to an alternative mode of civility, in which expressions of emotional spontaneity and otherworldly experience facilitated an improvement in personal and public morality. As Thomas Olivers argued, Methodist religiosity reformed popular culture, replacing oaths, curses and drinking songs with prayer, praise and sacred hymns.[94] According to another writer, it also substituted 'pretty polite Entertainments' with improving pursuits such as talking about the Spirit and listening to long, impassioned sermons.[95] The implication was that evangelical religion created a distinctively Christian soundscape which fostered morality without diminishing sociability. As James Kershaw wrote, 'They sing, they pray, and to each other speak, | And solely meet each others good to seek.'[96] In her poem 'On the Five Senses,' Margaret Davidson depicted hearing as the most important of the senses in Christian worship, for it facilitated communication with God. Unsurprisingly, sound had played a crucial role in Davidson's own conversion, when cries and groans were followed by the recitation of divinely inspired poetry. Moreover, after this passionate auditory experience, her ear became 'ravished' by heavenly music for a period of ten days, during which time her heart was weaned from the 'love of vain songs, and earthly musick.'[97] Instead of denying herself music, though, she rewrote her favourite tunes as hymns. Plainly, Methodists perceived the vocalisations, songs and conversations of evangelical devotion as crucial elements in a culture of self-improvement. In doing so, they heard their spirituality as a mode of civility rather than enthusiasm, even if they rarely described this loud, intense style of worship as noise. To be sure, Wesley and his followers found many different examples of auditory experience unacceptable. But despite some reservations, there was a consensus within Methodism that the sound of the Spirit constituted a legitimate source of improvement as well as inspiration, reform as well as revival.

Notes

1 Anon, *The Fanatic Saints; or, Bedlamites Inspired* (London: J. Bew, 1778), 8–10, 13, 20–23.
2 See Mark M. Smith, *Sensing the Past: Seeing, Hearing, Smelling, Tasting, and Touching in History* (Berkeley: University of California Press, 2007), esp. 4–18. See also Peter Denney, 'Looking Back, Groping Forward: Rethinking Sensory History,' *Rethinking History* 15, no. 4 (2011): 601–16; Martin Jay, 'In the Realm of the Senses: An Introduction,' *American Historical Review* 116, no. 2 (2011): 307–15; David Howes and Constance Classen, *Ways of Sensing: Understanding the Senses in Society* (London: Routledge, 2014), esp. 11–13.
3 Leigh Eric Schmidt, *Hearing Things: Religion, Illusion, and the American Enlightenment* (Cambridge, MA: Harvard University Press, 2000), 39.
4 Anne Taves, *Fits, Trances, and Visions: Experiencing Religion and Explaining Religious Experience from Wesley to James* (Princeton, NJ: Princeton University Press, 1999), 50–54.
5 Phyllis Mack, 'The Senses in Religion: Listening to God in the Eighteenth Century,' in *A Cultural History of the Senses in the Age of Enlightenment*, ed. Anne C. Vila (London: Bloomsbury, 2014), 94–97.
6 Theophilus Evans, *The History of Modern Enthusiasm* (London: W. Owen and W. Clarke, 1757), 128–29.
7 J.G.A Pocock, 'Clergy and Commerce: The Conservative Enlightenment in England,' in *L'Eta Dei Lumi*, ed. R. Ajello, E. Contese and V. Piano, 2 vols. (Naples: Jovene, 1985), 1:532–5; Lawrence E. Klein, 'Politeness and the Interpretation of the British Eighteenth Century,' *Historical Journal* 45, no. 4 (2002): 889–90.
8 Isabel Rivers, *Reason, Grace and Sentiment: A Study of the Language of Religion and Ethics in England, 1660–1780*, 2 vols. (Cambridge: Cambridge University Press, 1991), 1:36–37.
9 Edward Fowler, *The Design of Christianity*, 2nd ed. (London: R. Royston, 1675), 294–95.
10 Philip Carter, *Men and the Emergence of Polite Society, Britain 1660–1800* (Harlow: Pearson Education, 2001), 43.
11 John Hughes, *The Spectator*, no. 224 (1711), in Joseph Addison and Richard Steele, *The Spectator*, ed. G. Gregory Smith, 4 vols. (London: J. M. Dent, 1945), 2:167.

12 Richard Steele, *The Christian Hero* (London: J. Tonson, 1701), 41, 53. See also William Howdell, *Religion Productive of Joy, and Consistent with Politeness: A Sermon* (York: C. Ward, 1744), 23–24.
13 Joseph Addison, *Spectator*, no. 494 (1712), in Smith, *Spectator*, 4:64.
14 Joseph Addison, *Spectator*, no. 590 (1714), in Smith, *Spectator*, 4:351–52.
15 Howard D. Weinbrot, *Literature, Religion, and the Evolution of Culture, 1660–1780* (Baltimore, MD: Johns Hopkins University Press, 2013), 184–85.
16 Joseph Trapp, *The Nature, Usefulness, and Regulation of Religious Zeal: A Sermon* (London: L. Gilliver, 1739), 24, 27–28.
17 Trapp, *Nature, Usefulness, and Regulation of Religious Zeal*, 28.
18 Taves, *Fits, Trances and Visions*, 71–72.
19 David Hempton, *Methodism: Empire of the Spirit* (New Haven, CT: Yale University Press, 2005), 60–65.
20 Misty Anderson, *Imagining Methodism in Eighteenth-Century Britain: Enthusiasm, Belief and the Borders of the Self* (Baltimore, MD: Johns Hopkins University Press, 2012), 51–69.
21 Alexander Jephson, *A Friendly and Compassionate Address to all Serious and Well Disposed Methodists* (London: C. Jephson, 1760), 41–42.
22 Anon, *Fanatical Conversion; or, Methodism Displayed* (London: J. Bew, 1779), 15.
23 George Lavington, *The Enthusiasm of Methodists and Papists Compared*, 2 vols. (London: J. and P. Knapton, 1754), 2:93.
24 Jeremy Schmidt, *Melancholy and the Care of the Soul: Religion, Moral Philosophy and Madness in Early Modern England* (Aldershot: Ashgate, 2007), 150–62.
25 Jon Mee, *Romanticism, Enthusiasm and Regulation: Poetics and the Policing of Culture in the Romantic Period* (Oxford: Oxford University Press, 2003), 28–31.
26 Thomas Mortimer, *Die and be Damned*, 2nd ed. (London: S. Hooper and A. Morley, 1758), 42–43.
27 Hempton, *Methodism*, 56.
28 Anon, *Enthusiast, or Methodism Display'd*, 6.
29 James Adair, *The Methodist and Mimick* (London: C. Moran, 1766), 21. See also Anon, *Methodism and Popery Dissected and Compared* (London: Fielding and Walker, 1779), 31.
30 Leigh Schmidt, *Hearing Things*, 41–49.
31 Lavington, *Enthusiasm*, 2:68.
32 Lavington, *Enthusiasm*, 2:81–82.
33 William Penrice, *The Causes of Methodism set Forth* (London: no pub, 1771), 2.
34 John Mainwaring, *An Essay on the Character of Methodism* (Cambridge: J. Archdeacon, 1781), 24.
35 Brett C. McInelly, 'Raising the Roof: Hymn Singing, the Ant-Methodist Response, and Early Methodist Religiosity,' *Eighteenth-Century Life* 36, no. 2 (2012): 99–100.
36 Hempton, *Methodism*, 68–75.
37 Phyllis Mack, *Heart Religion in the British Enlightenment: Gender and Emotion in Early Methodism* (Cambridge: Cambridge University Press, 2008), 41–42.
38 Nicholas Temperley, 'John Wesley, Music, and the People Called Methodists,' in *Music and the Wesleys*, ed. Nicholas Temperley and Stephen Banfield (Urbana, IL: University of Illinois Press, 2010), 14–15.
39 William Vincent, *Considerations on Parochial Music* (London: T. Cadell, 1787), 13–14.

40 John Douglas, *The Criterion: or, Miracles Examined* (London: A. Millar, 1754), 175–76.
41 Lavington, *Enthusiasm*, 2:196.
42 Mee, *Romanticism, Regulation and Enthusiasm*, 29–30; Michael Heyd, *'Be Sober and Reasonable': The Critique of Enthusiasm in the Seventeenth and Early Eighteenth Centuries* (Leiden: Brill, 1995), 201–3.
43 Charles Avison, *An Essay on Musical Expression* (London: C. Davis, 1752), 3.
44 Thomas Green, *A Dissertation on Enthusiasm* (London: J. Oliver and T. Payne, 1755), 36.
45 Green, *Dissertation on Enthusiasm*, 118–19.
46 Graham J. Barker-Benfield, *The Culture of Sensibility: Sex and Society in Eighteenth-Century Britain* (Chicago: University of Chicago Press, 1992), 20–22; Shelley Trower, *Senses of Vibration: A History of the Pleasure and Pain of Sound* (New York: Continuum, 2012), 16–19.
47 John Stinstra, *A Pastoral Letter against Fanaticism* (London: A. Linde, 1753), 64.
48 John Scott, *A Fine Picture of Enthusiasm* (London: J. Noon, 1744), 27.
49 William Riley, *Parochial Music Corrected* (London: no pub, 1762), 8. See also Anon, *Of the Use and Abuse of Music* (London: no pub, 1753), 19.
50 Evans, *History of Modern Enthusiasm*, 132.
51 Charles Hoffer, *When Benjamin Franklin Met the Reverend Whitefield: Enlightenment, Revival, and the Power of the Printed Word* (Baltimore, MD: Johns Hopkins University Press, 2011), 17.
52 Harry S. Stout, *The Divine Dramatist: George Whitefield and the Rise of Modern Evangelicalism* (Grand Rapids: William B. Eerdmans, 1991), 40–44.
53 William Jay, *Memoirs of the Life and Character of Cornelius Winter* (Bath: M. Gye, 1808), 27–28.
54 Samuel Johnson and James Boswell, *A Journey to the Western Islands of Scotland* and *The Journal of a Tour to the Hebrides*, ed. Peter Levi (London: Penguin, 1984), 174. See also Anon, *Methodism and Enthusiasm Fully Display'd*, 2nd ed. (Oxford: M. Cooper, 1743), 14.
55 Thomas Chatterton, *The Poetical Works of Thomas Chatterton*, ed. Walter W. Skeat, 2 vols. (London: Bell and Daldy, 1871), 1:42.
56 *Gentleman's Magazine* 9 (1739), 242.
57 Joseph Trapp, *The Nature, Folly, Sin, and Danger, of Being Righteous Over-Much* (London: S. Austen, 1739), 57–58.
58 Green, *Dissertation on Enthusiasm*, 37–39.
59 Philagathus Cantabrigiensis, *The Methodist Instructed* (London: R. Withy, 1764), 7.
60 Albert M. Lyles, *Methodism Mocked: The Satiric Reaction to Methodism in the Eighteenth Century* (London: Epworth Press, 1960), 73–74.
61 Nathaniel Lancaster, *Methodism Triumphant, or, the Decisive Battle between the Old Servant and the Modern Saint* (London: J. Wilkie, 1767), 10.
62 John Mason, *An Essay on Elocution, or, Pronunciation* (London: M. Cooper, 1748), 6–7.
63 Paul Goring, 'Anglicanism, Enthusiasm and Quixotism: Preaching and Politeness in Mid-Eighteenth Century Literature,' *Literature and Theology* 15, no. 4 (2001): 331–34; Jennifer Farooq, *Preaching in Eighteenth-Century London* (Woodbridge: Boydell Press, 2013), 11–16.

64 David Fordyce, *Theodorus: A Dialogue Concerning the Art of Preaching* (London: R. Dodsley, 1752), 79–80.
65 Mortimer, *Die and be Damned*, 25–26. See also Alexander Stevens, *The Celebrated Lecture on Heads* (London: R. Bond, 1765), 19.
66 Brett C. McInelly, *Textual Warfare and the Making of Methodism* (Oxford: Oxford University Press, 2014), 5.
67 Hempton, *Methodism*, 33–41.
68 Anon, *Fanatical Conversion*, 28.
69 Bishop Gibson, *Observations upon the Conduct and Behaviour of a Certain Sect, Usually Distinguished by the Name of Methodists* (London: no pub, 1740), 10.
70 John Wesley, 'On Zeal,' in *Sermons on Several Occasions*, 8 vols. (London: J. Wesley, 1788), 7:263.
71 Wesley, 'On Pleasing all Men,' in *Sermons on Several Occasions*, 8:126.
72 John Wesley to Hannah Ball, 12 April 1774, in *The Letters of the Rev. John Wesley*, ed. John Telford, 8 vols. (1931; London: Epworth Press, 1960), 6:79.
73 John Wesley, *The Journal of the Rev. John Wesley*, ed. Ernest Rhys, 4 vols. (London: J. M. Dent, 1922), 3:37–38.
74 John Wesley to Charles Wesley, 11 December 1762, in *Letters of John Wesley*, 4:196.
75 John Walsh, 'Methodism and the Mob in the Eighteenth Century,' in *Popular Belief and Practice*, ed. G.J. Cuming and Derek Baker (Cambridge: Cambridge University Press, 1972), 222–24.
76 See Wesley, *Journal*, 1:271; *Journal*, 2:374.
77 Wesley, *Journal*, 3:139.
78 Wesley, *Journal*, 3:168.
79 Wesley, *Journal*, 2:224.
80 Wesley, *Journal*, 4:292.
81 Wesley, *Journal*, 1:408.
82 Wesley, *Journal*, 4:287.
83 Wesley, *Journal*, 3:153.
84 Wesley, *Journal*, 3:428.
85 Wesley, *Journal*, 3:253.
86 Taves, *Fits, Trances and Visions*, 56–58; Henry D. Rack, *Reasonable Enthusiast: John Wesley and the Rise if Methodism*, 3rd ed. (1989; London: Epworth Press, 2002), 276–77.
87 John Wesley, *A Farther Appeal to Men of Reason and Religion* (London: W. Strahan, 1745), 18.
88 Wesley, *Journal*, 2:489.
89 Wesley, *Journal*, 2:490.
90 Wesley, *Journal*, 3:449.
91 Wesley, *Journal*, 3:301.
92 Wesley, *Journal*, 4:139.
93 McInelly, *Textual Warfare*, 58.
94 Thomas Olivers, *A Defence of Methodism* (London: J. Atlay and R. Thackwray, 1785), 27, 46.
95 Anon, *Christianity Proved by Scripture Authority to be Neither Antinomianism nor Legality* (London: J. Fuller, 1758), 23.
96 James Kershaw, *The Methodist* (Nottingham: G. Burbage, 1780), 128.
97 Edward Smith, *The Extraordinary Life and Christian Experience of Margaret Davidson* (Dublin: B. Dugdale, 1782), 18.

Bibliography

Primary sources

Adair, James. *The Methodist and Mimick*. London: C. Moran, 1766.

Addison, Joseph and Richard Steele. *The Spectator* (1711–1714). Edited by G. Gregory Smith. 4 vols. London: J. M. Dent, 1945.

Anon. *Christianity Proved by Scripture Authority to be Neither Antinomianism nor Legality*. London: J. Fuller, 1758.

Anon. *The Fanatic Saints; or, Bedlamites Inspired*. London: J. Bew, 1778.

Anon. *Fanatical Conversion; or, Methodism Displayed*. London: J. Bew, 1779.

Anon. *Methodism and Enthusiasm Fully Display'd*. 2nd ed. Oxford: M. Cooper, 1743.

Anon. *Methodism and Popery Dissected and Compared*. London: Fielding and Walker, 1779.

Anon. *Of the Use and Abuse of Music*. London: no pub, 1753.

Avison, Charles. *An Essay on Musical Expression*. London: C. Davis, 1752.

Cantabrigiensis, Philagathus. *The Methodist Instructed*. London: R. Withy, 1764.

Chatterton, Thomas. *The Poetical Works of Thomas Chatterton*. Edited by Walter W. Skeat. 2 vols. London: Bell and Daldy, 1871.

Douglas, John. *The Criterion: Or, Miracles Examined*. London: A. Millar, 1754.

Evans, Theophilus. *The History of Modern Enthusiasm*. London: W. Owen and W. Clarke, 1757.

Fordyce, David. *Theodorus: A Dialogue Concerning the Art of Preaching*. London: R. Dodsley, 1752.

Fowler, Edward. *The Design of Christianity*. 2nd ed. London: R. Royston, 1675.

Gibson, Bishop. *Observations upon the Conduct and Behaviour of a Certain Sect, Usually Distinguished by the Name of Methodists*. London: no pub, 1740.

Green, Thomas. *A Dissertation on Enthusiasm*. London: J. Oliver and T. Payne, 1755.

Howdell, William. *Religion Productive of Joy, and Consistent with Politeness: A Sermon*. York: C. Ward, 1744.

Jay, William. *Memoirs of the Life and Character of Cornelius Winter*. Bath: M. Gye, 1808.

Jephson, Alexander. *A Friendly and Compassionate Address to all Serious and Well Disposed Methodists*. London: C. Jephson, 1760.

Johnson, Samuel and James Boswell. *A Journey to the Western Islands of Scotland* and *the Journal of a Tour to the Hebrides*. Edited by Peter Levi. London: Penguin, 1984.

Kershaw, James. *The Methodist*. Nottingham: G. Burbage, 1780.

Lancaster, Nathaniel. *Methodism Triumphant, or, the Decisive Battle between the Old Servant and the Modern Saint*. London: J. Wilkie, 1767.

Lavington, George. *The Enthusiasm of Methodists and Papists Compared*. 2 vols. London: J. and P. Knapton, 1754.

Mainwaring, John. *An Essay on the Character of Methodism*. Cambridge: J. Archdeacon, 1781.

Mason, John. *An Essay on Elocution, or, Pronunciation*. London: M. Cooper, 1748.

Mortimer, Thomas. *Die and be Damned*. 2nd ed. London: S. Hooper and A. Morley, 1758.
Olivers, Thomas. *A Defence of Methodism*. London: J. Atlay and R. Thackwray, 1785.
Penrice, William. *The Causes of Methodism Set Forth*. London: no pub, 1771.
Riley, William. *Parochial Music Corrected*. London: no pub, 1762.
Scott, John. *A Fine Picture of Enthusiasm*. London: J. Noon, 1744.
Smith, Edward. *The Extraordinary Life and Christian Experience of Margaret Davidson*. Dublin: B. Dugdale, 1782.
Steele, Richard. *The Christian Hero*. London: J. Tonson, 1701.
Stevens, Alexander. *The Celebrated Lecture on Heads*. London: R. Bond, 1765.
Stinstra, John. *A Pastoral Letter against Fanaticism*. London: A. Linde, 1753.
Trapp, Joseph. *The Nature, Folly, Sin, and Danger, of Being Righteous Over-Much*. London: S. Austen, 1739.
Trapp, Joseph. *The Nature, Usefulness, and Regulation of Religious Zeal: A Sermon*. London: L. Gilliver, 1739.
Vincent, William. *Considerations on Parochial Music*. London: T. Cadell, 1787.
Wesley, John. *A Farther Appeal to Men of Reason and Religion*. London: W. Strahan, 1745.
Wesley, John. *The Journal of the Rev. John Wesley*. Edited by Ernest Rhys. 4 vols. London: J. M. Dent, 1922.
Wesley, John. *The Letters of the Rev. John Wesley*. Edited by John Telford. 8 vols. 1931; London: Epworth Press, 1960.
Wesley, John. *Sermons on Several Occasions*. 8 vols. London: J. Wesley, 1788.

Secondary sources

Anderson, Misty. *Imagining Methodism in Eighteenth-Century Britain: Enthusiasm, Belief and the Borders of the Self*. Baltimore: Johns Hopkins University Press, 2012.
Barker-Benfield, Graham J. *The Culture of Sensibility: Sex and Society in Eighteenth-Century Britain*. Chicago: University of Chicago Press, 1992.
Carter, Philip. *Men and the Emergence of Polite Society, Britain 1660–1800*. Harlow: Pearson Education, 2001.
Denney, Peter. 'Looking Back, Groping Forward: Rethinking Sensory History.' *Rethinking History* 15, no. 4 (2011): 601–16.
Farooq, Jennifer. *Preaching in Eighteenth-Century London*. Woodbridge: Boydell Press, 2013.
Goring, Paul. 'Anglicanism, Enthusiasm and Quixotism: Preaching and Politeness in Mid-Eighteenth Century Literature.' *Literature and Theology* 15, no. 4 (2001): 326–41.
Hempton, David. *Methodism: Empire of the Spirit*. New Haven: Yale University Press, 2005.
Heyd, Michael. *'Be Sober and Reasonable': The Critique of Enthusiasm in the Seventeenth and Early Eighteenth Centuries*. Leiden: Brill, 1995.
Hoffer, Charles. *When Benjamin Franklin Met the Reverend Whitefield: Enlightenment, Revival, and the Power of the Printed Word*. Baltimore: Johns Hopkins University Press, 2011.
Howes, David and Constance Classen. *Ways of Sensing: Understanding the Senses in Society*. London: Routledge, 2014.

Jay, Martin. 'In the Realm of the Senses: An Introduction.' *American Historical Review* 116, no. 2 (2011): 307–15.

Klein, Lawrence E. 'Politeness and the Interpretation of the British Eighteenth Century.' *Historical Journal* 45, no. 4 (2002): 869–98.

Lyles, Albert M. *Methodism Mocked: The Satiric Reaction to Methodism in the Eighteenth Century*. London: Epworth Press, 1960.

Mack, Phyllis. 'The Senses in Religion: Listening to God in the Eighteenth Century.' In *A Cultural History of the Senses in the Age of Enlightenment*, edited by Anne C. Vila, 85–107. London: Bloomsbury, 2014.

Mack, Phyllis. *Heart Religion in the British Enlightenment: Gender and Emotion in Early Methodism*. Cambridge: Cambridge University Press, 2008.

McInelly, Brett C. 'Raising the Roof: Hymn Singing, the Ant-Methodist Response, and Early Methodist Religiosity.' *Eighteenth-Century Life* 36, no. 2 (2012): 80–110.

McInelly, Brett C. *Textual Warfare and the Making of Methodism*. Oxford: Oxford University Press, 2014.

Mee, Jon. *Romanticism, Enthusiasm and Regulation: Poetics and the Policing of Culture in the Romantic Period*. Oxford: Oxford University Press, 2003.

Pocock, J.G.A. 'Clergy and Commerce: The Conservative Enlightenment in England.' In *L'Eta Dei Lumi*, edited by R. Ajello, E. Contese and V. Piano, 1:523–62. 2 vols. Naples: Jovene, 1985.

Rack, Henry D. *Reasonable Enthusiast: John Wesley and the Rise if Methodism*. 3rd ed. 1989; London: Epworth Press, 2002.

Rivers, Isabel. *Reason, Grace and Sentiment: A Study of the Language of Religion and Ethics in England, 1660–1780*. 2 vols. Cambridge: Cambridge University Press, 1991.

Schmidt, Jeremy. *Melancholy and the Care of the Soul: Religion, Moral Philosophy and Madness in Early Modern England*. Aldershot: Ashgate, 2007.

Schmidt, Leigh Eric. *Hearing Things: Religion, Illusion, and the American Enlightenment*. Cambridge, MA: Harvard University Press, 2000.

Smith, Mark M. *Sensing the Past: Seeing, Hearing, Smelling, Tasting, and Touching in History*. Berkeley: University of California Press, 2007.

Stout, Harry S. *The Divine Dramatist: George Whitefield and the Rise of Modern Evangelicalism*. Grand Rapids: William B. Eerdmans, 1991.

Taves, Anne. *Fits, Trances, and Visions: Experiencing Religion and Explaining Religious Experience from Wesley to James*. Princeton: Princeton University Press, 1999.

Temperley, Nicholas. 'John Wesley, Music, and the People Called Methodists.' In *Music and the Wesleys*, edited by Nicholas Temperley and Stephen Banfield, 3–25. Urbana: University of Illinois Press, 2010.

Trower, Shelley. *Senses of Vibration: A History of the Pleasure and Pain of Sound*. New York: Continuum, 2012.

Walsh, John. 'Methodism and the Mob in the Eighteenth Century.' In *Popular Belief and Practice*, edited by G.J. Cuming and Derek Baker, 213–28. Cambridge: Cambridge University Press, 1972.

Weinbrot, Howard D. *Literature, Religion, and the Evolution of Culture, 1660–1780*. Baltimore: Johns Hopkins University Press, 2013.

7 Hissing the king

The politics of vocal expression in 1790s Britain

Gillian Russell

10 December 1806, the first night at Drury Lane theatre of a farce entitled *Mr. H*, was a humiliating experience for its author, Charles Lamb. He wrote later to a friend: 'Hang 'em, how they hissed! It was not a hiss neither, but a sort of frantic yell, like a congregation of mad geese, with a roaring sometimes like bears ... sometimes snakes, that hissed me into madness.'[1] Lamb's biographer Thomas Noon Talfourd claimed that Lamb went so far as to participate in his own damnation: giving way 'to the common feeling' he 'hissed and hooted as loudly as any of his neighbours.'[2] The custom of hissing, which a still wounded Lamb was to reflect on in an essay published five years later, was a frequently exercised form of audience expression in the Georgian theatre, used to indicate disapproval of a performance, a particular actor, or a perceived failure of theatre management to comply with the bill promised in newspaper or playbill publicity.[3] On some occasions, such dissent was politically motivated, as when the arrival of the Lord Lieutenant of Ireland at Smock Alley theatre in Dublin in July 1784 was greeted by 'chorusses of groaning, hissing, and shouting, with full accompanyments of claps, catcalls, whistles.'[4] In 1775, the notoriously litigious actor Charles Macklin brought a case against two men, James Sparks and Samuel Reddish, whose hissing, he believed, had led to the failure of his production of *Macbeth* at Covent Garden theatre and his eventual sacking by the manager George Colman in 1773. During the trial, the Chief Justice Lord Mansfield defined 'hissing and applauding' as an 'unalterable right' of audiences, but he distinguished between such behaviour as a 'natural' spontaneous response and as premeditated intimidation designed to rob an individual of his livelihood.[5]

The expression of dissent in the theatre through sounds such as hissing was a long-standing practice that can be traced back to antiquity. In *The Laws*, Plato made such vociferousness central to an argument about the politics of theatre and, conversely, the theatre of politics. Using the figure of the Athenian to dialogise his case, Plato claimed that the adulteration of musical forms had licensed a dangerous tendency in audiences to voice their opinions through 'catcalls and discordant outcries of the crowd.' Audiences that were previously quiet and disciplined were now vocal and noisy: 'our once silent audiences have found a voice, in the persuasion that they understand what is good and bad in art; the old sovereignty of the best, aristocracy, has given way to an evil "sovereignty of the audience," a theatrocracy.'[6] The latter, moreover, was more dangerous than a democracy because it was based on the universal communicability of voices and was thus potentially inclusive of slaves and women. Another manifestation of this theatrocracy was disrespect for the sanctity of the performance space: as evidence of this, Plato's Athenian cites the tendency for itinerant choirs to interrupt ceremonies by turning up and staging their own performances in close proximity to the altar.[7] As Samuel Weber argues, theatrocracy's

> true resource ... consists in the power to *move* and disrupt the consecrated and institutionalized boundaries that structure political space: those, for instance, that separate the sacred from the profane, the 'altar' from the public. Theatricality demonstrates its subversive power when it forsakes the confines of the *theatron* and begins to wander: when, in short, it separates itself from the theater.[8]

In late-eighteenth-century Britain, the tendency of theatre audiences to sound both approval and dissent had become a norm, an established prerogative, as Lord Mansfield had acknowledged when he described hissing as an 'unalterable right.' Practices such as hissing and the catcall were part of a repertoire of sonic effects by which the audience manifested itself as a social body and asserted its rights before actors, theatre managers and society as a whole. Such performances could also be for the audience's own entertainment as well as affirming distinctions within the auditorium itself. The sonic effects of theatre audiences included laughter, sobbing, the clapping of hands, the stamping of feet and general talk, making theatre buildings, especially wooden ones, sound boxes for sociality.[9] A notable example of the latter is the playwright Frederick Reynolds's account of how, as a child, he happened to pass by Drury Lane theatre on the first night of Richard Brinsley Sheridan's *The School for Scandal* in 1777. Hearing

> such a tremendous noise over my head [and] fearing the Theatre was proceeding to fall about it, I ran for my life; but found, the next morning, that the noise did not arise from the *falling* of the house, but from the *falling* of the screen in the fourth act, so violent, and so tumultuous were the applause and laughter.[10]

Hissing, as Charles Lamb discovered, could be similarly 'violent' and overwhelming. Associated with baring the teeth in a hostile or defensive gesture, and linked in comments such as Lamb's with snakes or geese, the origins of hissing probably go far back into human history. Hissing is produced by moving the tongue close to the roof of the mouth and pushing air past it and through the teeth, in a similar manner to the production of fricative consonants such as 's' and 'f.' It does not entail the exercise of the vocal chords, like the sigh, though hissing differs from sighing or coughing in never being involuntary. As Steven Connor argues in his account of non-lexical utterances, the sound signified by the aspirate consonant 'h' 'approaches ... the degree zero of consonance ... [belonging] wholly neither to voice nor to noise, neither to larynx nor to mouth.'[11] Insofar that the 'h' sound in hiss is often barely audible, the hiss is even more indeterminate in its origins, being neither vocalic nor properly consonantal. Nonetheless, while a form of human 'noise,' hissing is also a kind of voice because it is voluntary: one hisses through instinctual agency rather than because of a physical reflex. Hissing as a form of noise/voice is thus highly performative, simultaneously an expression of advanced sociality as well as always being latently primitive, even animalistic. If, as Connor argues, 'the secret of *h* is that it is the ultimate mixed body, the Hermes of the crossroads where vowel and consonant, voice and void, sound and sense, soul and machine, meet,' then the hiss is a collectively expressed *h*, the ultimate expression of the sovereignty of the 'mixed body' of Plato's theatrocracy in its capacity to disrupt the performance and extend the boundaries of the audience.[12]

In spite of its long history, however, hissing and other related forms of audience noisemaking such as the catcall have received very little attention in theatre history.[13] Nor has the relationship of such sounds to communal vocal expression outside the theatre been significantly addressed. Hissing was part of the often highly ritualised repertoire of Georgian crowd behaviour, exercised on occasions such as riots as well as elections and ceremonial events. One advantage of hissing was that, unlike excitable gesticulating, stone-throwing, and shouting, it could be done in a way that preserved corporeal control and decorum. A hissing individual could be static, otherwise expressionless, making hissing a disciplined performance in its own right, and thereby potentially more threatening. In a crowd, it would have been difficult to determine who started hissing unless you were standing next to the particular person or persons concerned, and even then there might still be some uncertainty as to from whom a hiss had emanated. Once hissing began, its controllability was as uncertain as wildfire, its modulation ranging from the faintest sibilance to what Charles Lamb in his 1811 essay evocatively described as 'like the noise of many waters.'[14] The hiss was therefore one of the most powerful expressions of the collective voice in late Georgian culture. As a form of vocal behaviour that could be heard on the streets and in venues such as theatres, it crossed an important ideological and material boundary in cultural and political life: between the extra-parliamentary sphere of the people 'out-of-doors' and the 'indoors' of social, cultural and political power.

In accounts of such behaviour, hissing is often linked with the practices of hooting and groaning, as in the description of the response to the Lord Lieutenant in Smock Alley theatre. Hooting is an inarticulate shout or exclamation, sometimes merging with the yell or whoop, while the groan is a low vocal murmur, traditionally associated with religious humility or abjection. Groaning was linked with sighing as involuntary expressions of the operation in the body of the Holy Spirit.[15] In secular contexts, groaning could be a gesture of reflexive objectification, an expression of collective suffering that demanded redress or restitution. Hooting and groaning were thus potentially ambiguous in the way that a hiss never was: they could signify, in the case of a hoot, either approbation or disapprobation, or, in the case of groaning, possibly passive, non-violent supplication.

An example of the conjunction of hissing, groaning and hooting in this period is an incident from the 1830s, reported in *The Examiner.* Michael Shields, a porter at Covent Garden, had attempted to return to his work after he was discharged in a case of bodysnatching. On his return to the market, he was 'almost immediately assailed by a general groan and other demonstrations of indignation and disgust.' A crowd of approximately 100 then pursued him through the streets around Covent Garden 'groaning, hooting, and hissing him as he ran.' Shields eventually sought sanctuary at the police station in Bow Street, where the crowd 'continued for an hour in front of the office … hissing and groaning, and calling upon the magistrates and officers to turn the old man into the street.'[16] Shields was eventually escorted to safety after the crowd dispersed. His ostracising from the community in which he lived and worked had been performatively achieved through the repertoire of hissing, groaning and hooting, and the crowd's use of movement and space in their pursuit of him through the London streets.

The mainly sonic attack on Shields is an example of how such behaviour could constitute a micropolitical event, exemplifying a quotidian and ephemeral politics of the street. We know of this incident because it was recorded in the press, but there are probably many others, momentary but expressive, that escaped notice. The repertoire of hissing, groaning and hooting, for example, was part of the characteristic behaviour of crowds in late Georgian food riots, which are at the centre of E.P. Thompson's influential concept of the 'moral economy' of the crowd, that is, the assertion of rights other than the profit of the seller as part of an economic transaction, such as, in the case of bread, the right of the poor to be able to obtain it at a price they could afford.[17] The role of hissing in food riots can be illustrated by the disturbances that took place in London between 15 and 18 September 1800, following agitation against food shortages during the summer of that year in Yorkshire, Nottinghamshire, the West Midlands and Oxfordshire.[18] Disturbances in the metropolis (which also extended to other areas of the south-east, including Kent and Hampshire) focussed on the Corn Exchange in Mark Lane in the City of London. They were triggered, according to newspaper reports, by the strategic deployment on Sunday 14 September of ephemeral print, large written posters stuck up around London Bridge and the Monument, urging the poor to protest against the high price of bread, that were combined with smaller printed handbills. On the following Monday, large crowds of between 1,000 and 2,000 people assembled in Mark Lane, shouting slogans such as 'bread, bread, give us bread, don't starve us'; according to the *Courier* newspaper, 'no acts of violence were committed, and the mob confined themselves to hissing.'[19] A particular target of the crowd's reprobation were Quakers who were prominent in the corn trade: 'the populace bestowed on them some marks of their antipathy and vengeance, by hissing and hooting, and pelting them with mud.'[20] The disturbances went on for three days, Roger Wells noting that the stipendiary magistrates were 'pleasantly surprised that … incidents were restricted to expressions of "particular resentment" against the more notorious traders, to be distinguished from any "general disposition to riot and disturbance".'[21] The sounds of hissing and hooting were the main way that this 'resentment' was expressed: in addition to the Quakers, other dealers in corn were subject to 'hissing and groaning,' as were the Volunteers when they were deployed on Monday 15 September. Even the Lord Mayor was 'groaned and hissed in the most tumultuous manner. Every mouth seemed opened, and the neighbouring streets re-echoed with the cries of "*Bread! Bread!*"'[22] The Riot Act was read on Monday 15 September, but on Wednesday the 17th, the Lord Mayor declared that it would be more strictly enforced, causing the protesters to change tactics. They resorted to moving quickly through the streets in order to avoid the authorities, smashing street lamps as they went, in what Wells describes as 'a disciplined and extensive itineracy,' similar to that practised by the pursuers of Shields 30 years later and resonating with Weber's emphasis on theatrocracy's 'true resource' as the refusal to be confined to a particular space, to 'wander' and confound the power invested in the theatron.[23] As one commentator on the Mark Lane riots noted, 'no one could tell where the mob was, but every one knew where they had been.'[24]

Hissing, the related practices of hooting and groaning, and the movement of protesters through the streets, refusing to be static and thereby evading the performative reading of the Riot Act, were essential to the disturbances of September 1800 as a struggle for moral, political and physical territory and, insofar as many of the participants were hungry, Agambenian 'bare life.' As Thompson notes, 'it is restraint, rather than disorder, which is remarkable' in the 'moral economy' of the bread riot, and the vocality of the crowd was crucial to the discipline and theatricality of that 'restraint,' enabling the protest to articulate grievances but at the same time stay within the boundaries of legality.[25] Insofar as it could function in this way, hissing was thus part of the negotiation of civility in the broadest sense, a fact reinforced by its prevalence in the theatre and in other forums of polite sociability. Hissing was a practice that was not confined to plebeians. It crossed class boundaries, as indeed did the sound itself, in that hissing was not inflected by gender, class or regional accents in the way that words were: the sound alone could not 'place' its speaker but needed context to be meaningful. Hissing was thus a potentially universal language of dissent in eighteenth-century culture, a highly ritualised and potent weapon in all forms of 'the mixed body' of communal sociability, both polite and impolite, plebeian and elite. The gentleman musician and diarist John Marsh, for example, reports a number of occasions on which hissing occurred in the genteel context of music concerts in assembly rooms, including during a performance of Handel's *Messiah*:

> it happen'd once at a concert about Easter [in Salisbury in 1780] where the latter part of the Messiah was perform'd that there was no bass singer present to sing the trumpet song, on w'ch [a singer] thought proper to attempt it for w'ch he got handsomely hiss'd.[26]

Hissing was also customary in the metropolitan opera houses, as when, in 1795, a 'herd of powdered puppies' in the pit of the King's Theatre objected to the performance of a ballet by 'a considerable clamour, hissing.'[27] Hissing in such contexts tested the limits of politeness, enacting the boundaries of what would or would not be tolerated by audiences, in a way that was crucial to the maintenance of 'proper' sociability; conversely, hissing enabled plebeian audiences, both within and without the theatre, to exercise the universal voice of dissent and thus affirm the legitimacy of their place within the mixed social body as a whole.

The uncertainty as to whether the hiss was polite or not, whether it signalled passive or active opposition, 'restraint' or violence, civility or anarchy, came to the fore in the mid-1790s, in a series of events that highlight the intimacy between the theatre as a cultural institution and the theatricality of politics, as well as the crisis of these years as a crisis of theatrocracy. Plato's emphasis on the threat to the stability of the social order as a whole, represented by a noisy, restless audience, resonates with responses to the French Revolution such as Edmund Burke's characterisation of the National Assembly as

> acting like comedians of a fair before a riotous audience; they act amidst the tumultuous cries of a mixed mob of ferocious men, and of women lost to shame, who, according to their insolent fancies, direct, control, applaud, explode them ... As they have inverted order in all things, the gallery is in the place of the house.[28]

After 1789, the meaning of an assembly of people out of doors, whether as a protest meeting, as part of an election crowd, a bread riot, a ceremonial royal occasion, celebrations of wartime victories or gatherings at outdoor fairs such as Bartholomew Fair, was complicated by the example of the apparently successful usurpation in France of the sovereignty of the monarch by the sovereignty of the people, exemplified by the crowd spectacles of the Revolutionary Festivals.[29] In Britain, customs of theatre and counter-theatre by which, in Thompson's well-known formulation, rulers and ruled performed for each other in a way that maintained the show of paternalistic social relations were being enacted in different circumstances in the 1790s, intensifying the politics of voice in this decade.[30] In the light of the instantiation in France of the sovereignty of the people, what did it mean to evoke the people's voice through oratorical performances (in all the 'theatres' of the polity, such as parliament, the open field, the tavern, the lecture room, and the theatre itself); what, moreover, if the people spoke for themselves in the voice of sovereignty? As the 'ultimate mixed body,' in Connor's terms, of noise/voice, the hiss was highly significant to a 1790s politics of voice, not only for its relevance to the idea of theatrocracy as a form of dangerous adulteration, or mixing, of the proper social order (evident in Burke's reference to the 'mixed mob'), but also because it highlights another aspect of Plato's concern about theatrocracy – how it derived from the dilution of the hierarchy of genres in musical performances, leading to 'a confusion of forms' that conceded authority over the regulation of taste to the voice of the ever-restive audience. As Weber notes, the 'theatrocratic usurpation of the rule of law' is predominantly 'acoustical' rather than 'visual': 'song, dance, and music break down most effectively the sense of propriety and the barriers that are [theatrocracy's] condition, giving the "silent" majority a voice and producing a hybrid music bordering on noise.' 'The emergence of theatrocracy,' Weber goes on to state, 'thus necessarily involves what today we would call "multimedia".'[31] The theatrocratic crisis of the 1790s has a similar dual dimension in that it is concerned with both the challenge of the 'noise' of the people to the stability of the polity and a concomitant adulteration of genres and media. One particular dimension of the latter was the explosion in the 1790s of the ephemeral print forms such as handbills that were produced to the moment and served to amplify as well as commemorate the evanescence of public utterance.[32] An example of the uses of such media is the deployment of posters and handbills in the Mark Lane disturbances of 1800. While the publication of songs in the 1790s is clearly evidence of the synergies of print and the acoustic world, other kinds of ephemeral print, I would argue, played a similar kind of intermedial role.

The importance of the hiss to the politics of voice in the 1790s is illustrated by a series of events that occurred in London on 29 and

30 October 1795, the culmination of agitation for political reform led by the London Corresponding Society (LCS) that began with a mass meeting at St George's Fields in Southwark on 29 June 1795.[33] The June meeting was not truly 'open' in so far that the LCS issued tickets at 6d. each to a central area of the field, enclosed by a paling, where speeches were made that attempted to capitalise on opposition to the war with France and increased food prices and shortages.[34] The LCS received no response over the summer to an address to the King appealing for 'a timely attention to the Cries of a suffering People ... to give them ... free and equal Representation' and another larger meeting was held on 26 October 1795 at Copenhagen Fields in Islington, with possibly 100,000 in attendance.[35] On this occasion, there were no limitations on access: it was the most openly public gathering that the LCS had staged so far. The LCS pamphlet produced in conjunction with the meeting acknowledged the 'danger of convening large assemblies of the people' but noted that the peaceable behaviour of the crowd on this day gave the lie to assumptions of inevitable 'tumult and disorder.'[36] The crowd behaved with acoustic decorum, receiving the LCS addresses with 'the warmest and most unanimous acclamations of approbation,' while the *Gentleman's Magazine*, no friend to radical politics, reported that 'the multitude dispersed in the utmost quietness.'[37] As was the case at the June meeting, the centrepiece of the assembly at Copenhagen Fields was an address to George III that in this case took the form of a 'Remonstrance,' demanding that the King listen to 'the cries of a wearied and afflicted people.'[38] The use of the term 'remonstrance' had more performative force than 'address' or 'petition.' Recalling the 1641 Grand Remonstrance of Parliament to Charles I that led to the English Civil War, it suggested the legitimate right to collective protest of this 'parliament' of the people, and the necessity that George III fulfil his coronation oath by attending to their cries. As John Barrell notes, both the Copenhagen Fields addresses of October 1795 and those of George's Fields meeting of June

> contained ... an indefinite threat to the king's life; an imagining of a time, which could no longer be deferred, when ... popular resentment would be so impatient as to endanger the throne. "It is our right," the "Remonstrance" concluded, "to advise as well as supplicate".[39]

Three days after the Copenhagen Fields meeting, George III was scheduled to give his speech at the state opening parliament, which was widely rumoured beforehand to offer no alleviation of the privations caused by the war. The events of 29 October were reported in two anonymous pamphlets published soon after: the loyalist *Narrative of the Insults Offered to the King*, 'By an Eye-Witness' and *Truth and Treason!*, a penny pamphlet, which signalled its radical politics by means of parodic imitation of the overblown loyalist encomia to the monarch.[40] According to *Insults Offered to the King*, people began gathering on the route from St James's Palace to Westminster early on the morning of the 29th:

> Small parties, consisting of ten or twelve persons, mean and dirty in their habits, with gloomy countenances ... were seen coming from all parts of the town towards St James and Westminster. Silent and solemn they marched along, their heads, perhaps, full of the instructions they had received, or perhaps pondering upon the rewards they had been promised.[41]

The silence of crowds could be as meaningful as vociferousness, suggesting in this case that the people had been dangerously suborned by the LCS. By mid-afternoon, the crowd had grown to significant proportions, both pamphlets estimating it at 200,000. At 2.15, the King left St James's Palace in the State Coach. According to *Truth and Treason!,* as it entered the Mall, 'the multitude assailed him ... with Hisses and Groans, and other marks of Disaffection,' an account confirmed by *Insults Offered to the King* which reported that when the carriage appeared the crowd 'began to hiss and groan.'[42] The gesture of groaning, as an expression of supplication to the monarch, amplified the effect of the 'remonstrance' of the Copenhagen Fields meeting of just a few days before. Both pamphlets reported cries of 'No War,' 'Bread! Bread!' and, in line with the ritual theatricality of food riots, a number of people held up small loaves of bread 'decorated with black crape' to signify dearth.[43]

The actions of the people, an assembly that was legitimated by the state itself in the form of the ceremonial procession for the state opening of parliament, therefore combined a number of what might be called 'genres' of crowd behaviour – the food 'riot,' an anti-war demonstration and a continuation of the LCS meeting of 26 October. The fact that George III's display of himself involved movement through the streets licensed the crowd's theatrocratic behaviour in the form of the exercising of its voice and a dynamic itinerancy, the events of these two days exemplifying, in Weber's terms, 'the power to *move* and disrupt the consecrated and institutionalized boundaries that structure political space.' Hence, the crowd assembled to view George III was not static but followed the State Coach as it crossed Horse Guards into Whitehall and Parliament Street 'in the midst of hooting, hissing, and execrations.'[44] Some of the glasses of the State Coach were broken, and in Old Palace Yard, a missile, which both pamphlets described as a ball from an air gun, penetrated one of the side windows. *Truth and Treason!* reported in mock horror that it was 'intended no doubt to assassinate our *Sacred and Dread Sovereign.*'[45] George III indeed believed that he had been targeted, declaring to the Lord Chancellor with the firmness conveyed by what Barrell terms the report's 'outraged capitals,' 'MY LORD, I HAVE

BEEN SHOT AT!' or with less resolution, the stuttering speech patterns reported in *Truth and Treason!* 'My Lord, I, I, I've been shot at!!'[46] At three o'clock, the King left Westminster to the sound of ceremonial guns, the populace, according to *Truth and Treason!*, greeting him with 'a most hideous yell, almost as horrible as the roaring of *Aetna* ...'[47] *A Narrative of the Insults* claimed that 'Jacobinical and Corresponding agents,' no more than 50 or 60 in number, were active in leading the chorus of 'cries, hisses and groans' that followed the King on his return to St James's, a chorus which was 'the signal for others in more distant situations to repeat them.'[48] On returning to the Royal Mews empty of the monarch after delivering him to St James's Palace, the State Coach was surrounded by a crowd who proceeded to beat in the panels and collect pieces of broken windows as souvenirs.[49]

In spite of what had happened, in an act of either brazen courage or foolishness, at four o'clock the King decided to venture out once more, travelling to Buckingham House in a private coach with an escort of only two footmen. Once more he was greeted by hissing, hooting and a barrage of stones. The coach was surrounded, and a man opened the door with the apparent aim of dragging out the King before the Horse Guards belatedly arrived to give protection.[50] The following evening, Friday 30 October, the King and the royal family attended Covent Garden theatre in a demonstration of royal fortitude, with dragoons lining the streets to protect them. People continued to sound their disapproval, following the carriages, according to *A Narrative of the Insults*, with 'hisses and execrations.'[51] Even within Covent Garden theatre, which the royal family had chosen because it was regarded as more loyal than Drury Lane, the de facto headquarters of Sheridan's Whigs, some members of the audience dared to hiss, until, according to *Truth and Treason!*, 'their rebellious noises were drowned, by the loyal shoutings' of the King's supporters ... [M]any of the hissers were knocked down, and thrust out of the house.'[52] *A Narrative of the Insults* also acknowledged the 'few solitary hisses' to be heard in the theatre, issuing 'from a dozen or two of democratic serpents.'[53] Both pamphlets reprinted newspaper reports of the King's appearance at Covent Garden, including one extolling the 'acclamations' of the evening as 'like the balm of Gilead to heal the wounded spirit of royalty.'[54] As the royal family were leaving the theatre, the Covent Garden orchestra played the Duke of York's march, 'the Audience beating time with clapping of hands, vehement, but regular.'[55] The hisses that had assailed George III over the course of two days were being resisted by a form of loyalist sonic counter-theatre, created by bodies in militaristic 'regular' synchronicity. The sound created was as much for the benefit of those outside the theatre as for those within. Frederick Reynolds's boyhood experience showed that theatre walls could be permeable, the theatre having a privileged status as a sonic arena that in this case linked the political domain of 'indoors' with the disenfranchised 'out-of-doors.'

The possibility of an assassination attempt on the King has understandably dominated analysis of the events of 29 and 30 October. John Barrell has described the incident in Old Palace Yard as 'the answer to a loyalist prayer,' used to implement quickly the Two Acts, consisting of the Seditious Meetings Act, designed to restrict political lectures and meetings and mass gatherings such as the one in Copenhagen Fields, and the Treasonable Practices Act.[56] The latter refined an act of Edward III to make conspiracy to imagine or conceive the King's death a substantive treason. To 'express, utter, or declare, by any printing, writing, preaching, or malicious and advised speaking,' a threat to the person of the King was deemed evidence of traitorous intent, a formulation that invests much in the performative power of the word. Print is subsumed within a conception of utterance that blurs the boundaries between speech and the writing.[57] Although the possibility that the King had been shot at was undoubtedly the catalyst for the subsequent crackdown on rights of assembly and political expression, the sustained vocal attack that George III received over two days was also important and in fact dominates the accounts in the two pamphlets I have been citing. The events of 29 and 30 October were characterised by myriad, mutually resonating speech acts, both formal and informal: the reported hissing, hooting and groaning of the crowd, the exclamations of a shaken George III, the speech he gave in parliament, the subsequent proclamations read out to the populace, the performances of the actors in Covent Garden theatre and the 'shoutings' of the audience and so on. This texture of sounds was mapped onto and constitutes a dimension of the event's spatiality: as *A Narrative of the Insults* suggested, hissing created a political soundscape, serving as a signal to those in more 'distant situations' to move into the very centre of British power. The mass meetings of June and October 1795 had taken place on the margins of the city and Westminster, in Islington and south of the river in Southwark. The events of 29–30 October represented an incursion of the democratic crowd into royal ceremonial space and, moreover, a refusal to play the part of deferential spectator in the theatre of royalty, thereby denying the legitimacy of the King's authority in his person. By refusing to play their natural part, disrupting the procession, chasing the King's coach and even dismantling it, the crowd staged a form of counter-theatrical resistance that would not allow the performance of monarchical power to cohere. This resistance was manifested most clearly, not just by physical violence, but by the repertoire of the collective voice – hissing, hooting and groaning – which signified both disapproval of the show and the customary right of the people to express their disapproval, a right to their own concrete existence and embodiment as a public.

The crisis of the 1790s has tended to be viewed through the lens of written discourse, particularly print, but attentiveness to the voice suggests other possible ways of interpreting that crisis, and also other ways of defining the role of print itself. An important dimension of those two days in October 1795 was the material presence of fugitive or ephemeral print – single-sheet publications of all kinds. *Truth and Treason!* drew attention, in tones of mock disdain, to the '*vile miscreant* near the Palace Gate, with a breat [sic] bundle of Books and Papers, crying out, "*The Rights of Man for a Penny*",' while *A Narrative of the Insults* singled out a seditious handbill, entitled (again with outraged capitals) 'KILLING KINGS NO MURDER.'[58] This reference to a handbill, with the actual title of *King Killing* and printed and distributed at the Copenhagen Fields meeting by Richard 'Citizen' Lee, was designed to suggest that the attack on the King three days later was a premeditated act of the LCS. The author of *A Narrative of the Insults* cited this 'half sheet tract' – it was important that his readers were aware of the material form of the publication and thus its low status – as instrumental to the 'concatenation' of speech acts, such as public lectures, field meetings and more sinister 'private conclaves' which created a 'chain of sedition through the land.'[59] Cried out or performed, as part of the events and the ephemerality of a particular day, these texts were part of the political soundscape of hissing, hooting and groaning. They were a form of print to the moment, designed to emulate and reinforce the communicative power of voices crying to voices, and of bodies moving in pursuit of a king through the streets. Royal power was similarly invested in this form of printed utterance, the main form of communication with the King's subjects being through proclamations, ceremonially enunciated and then pasted up on posts and walls. The performances at the two royal theatres of Covent Garden and Drury Lane were also publicised via playbills, cried up and sold in the streets outside the playhouses. Ephemeral print of all kinds and the manifold political voice were therefore mutually implicated as utterances to the moment.

The print trade was also directly involved in the events of 29–30 October, generating a range of political actors. One of the constables protecting the King's body in the state coach was the loyalist bookseller John Stockdale of Piccadilly who was one of the chief witnesses of the possible assassination attempt. Stockdale later gave evidence in the trial of Kidd Wake, who as a journeyman printer also made his living from the press. Wake was tried in February 1796 for hissing, hooting, making 'wry faces' and uttering scandalous words against the King.[60] The government had been unable to charge anyone with an assassination attempt against George III even though a £1,000 reward had been offered for information, nor had they been able to implicate the LCS directly, as neither Wake nor Edward Collins, who was also arrested for stoning the state coach, were LCS members. Wake's punishment was a period of three months of daily public exposure in the pillory and five years of hard labour in Gloucester gaol. Acting in his defence, Thomas Erskine tried to ameliorate the severity of the sentence by claiming that Wake made faces at the King because he was short-sighted and was trying to see his monarch properly. The ludicrousness of Erskine's argument draws attention to how forms of political protest in the eighteenth century could be intimate and particular, an eyeball-to-eyeball confrontation, nullifying the distance between ruler and ruled. The violence of the loss of monarchical aura in a recognition of mutual humanity was potentially as searing as a physical wounding.

As a form of communication, hissing was adaptable to both collective protest and such intimate, personal encounters. In a pamphlet published after his release from prison in 1801, Kidd Wake defended his right to hiss, claiming that he was agitating for an end to the war rather than the potentially treasonous downfall of the King:

> I do confess that, ardently wishing for peace, and animated by the general clamour for that object, I was hurried beyond discretion or decency – I did join in the hissing, and in the cry of *No War*; but in no other whatever. My idea, at the moment was, that his Majesty would see how much the people disapproved the war, and that he might attend to their voice.[61]

For Kidd Wake and many others, to be a British subject in 1795 was not only a matter of access to individual free speech but also to a right to express the communal voice, to be a member of an audience in other words who could hiss if he (or she) did not like the show.

Attending to the voice, to its many forms, the sound textures it creates, and how ephemeral forms of printed utterance are mutually implicated with it, complicates our understanding of late Georgian civility, particularly in relation to how forms of civility out-of-doors intersect with those indoors. The repertoire of hissing, hooting and groaning continued as part of crowd behaviour well into the nineteenth century, as a feature of political protest, food riots and royal ceremonial occasions, such as the coronation of George IV in 1821. It also endured as a customary expression of audience dissent in the theatre, most notably in the Old Price riots of 1809.[62] The reasons for the decline of hissing are beyond the scope of this chapter, but by the 1860s, it was being observed that 'the old salutary practice of hissing in the theatres was falling into disuse.'[63] Hissing was gradually replaced by booing, which survives as an expression of audience disapproval and recalcitrant independence in the theatre, opera house and sporting arena.[64] The events of October 1795 highlight that the gathering of bodies and the articulation of voices in ephemeral time and space are not necessarily pre- or parapolitical – 'real' politics being something that happens elsewhere, beyond the reach of these voices – but can be seen as forms of performative collectivity that actualised in real time what was at stake in the confrontation between a king and his people. As Judith Butler argues in relation to contemporary twenty-first century conditions of the possibilities of dissent,

> [e]mbodied actions of various kinds signify in ways that are, strictly speaking, neither discursive or prediscursive. In other words forms of assembly already signify prior to, and apart from, any particular demands they make.... [T]he power that people have to gather together is itself an important political prerogative, quite distinct from the right to say whatever they have to say once people have gathered. The gathering signifies in excess of what is said, and that mode of signification is a concerted bodily enactment, a plural form of performativity.[65]

The events of the 1790s suggest that there is a long history to this kind of plural performativity, the hiss and other similar forms of voice/noise representing 'bodily enactment' as a form of saying, occupying the hyphenated space between the discursive and the prediscursive. The theatricality of hissing also meant that such forms of 'bodily enactment' were not truly ephemeral but could be commemorated and repeated to form a kind of archive: in opening their many mouths as one, the participants in the Mark Lane food riots of 1800 were not only sounding their own grievances, but also echoing the cries of 1795.

Postscript

The vestigial power of that archive can be discerned as recently as December 2010, when Prince Charles and the Duchess of Cornwall, en route to the Royal Variety Performance at the London Palladium, were confronted with students protesting at the introduction of tuition fees in UK higher education. A YouTube video, a twenty-first century equivalent of *Truth and Treason!*, shows students pursuing the royal car through the streets, hooting and practising a form of groaning, if not clearly hissing.[66] The response of the royal couple, a mixture of bafflement and fear, echoes George III's incomprehension in October 1795. A contemporary instance of theatrocratic politics, the YouTube video illustrates how the repertoire of protest from the 1790s continues to carry or bear meaning and how, like Kidd Wake in 1795, the 'people' are still demanding that those in state 'attend' to their voice.

Notes

1 Charles Lamb, *The Letters of Charles Lamb, with a Sketch of his Life by Thomas Noon Talfourd*, 2 vols. (London: Edward Moxon, 1837), 1:303.
2 Lamb, *Letters*, 1:297.
3 Charles Lamb, 'On the Custom of Hissing at the Theatres, with some Account of a Club of Damned Authors,' *The Reflector* 2 (1811): 122–27.
4 *Morning Herald*, 19 July 1784. For the contexts of this incident, see Helen Burke, 'Acting in the Periphery: The Irish Theatre,' in *The Cambridge*

Companion to British Theatre, 1730–1830, ed. Jane Moody and Daniel O'Quinn (Cambridge: Cambridge University Press, 2007), 227–28.

5 Francis Aspry Congreve, *Authentic Memoirs of the Late Mr. Charles Macklin, Comedian* (London: J. Barker, 1798), 53.

6 Plato, *The Laws*, trans. A.E. Taylor, 700 a-d, in Plato, *The Collected Dialogues*, ed. Edith Hamilton and Huntington Cairns (Princeton, NJ: Princeton University Press, 1973), 1294. For an application of this argument to song cultures of the 1790s to which I am indebted, see Kate Horgan, *The Politics of Songs in Eighteenth-Century Britain, 1723–1795* (Abingdon: Pickering and Chatto, 2014).

7 Plato, *The Laws*, 1372.

8 Samuel Weber, *Theatricality as Medium* (New York, NY: Fordham University Press, 2004), 37.

9 For an influential account of the early modern theatre as a sonic 'instrument,' see Bruce R. Smith, *The Acoustic World of Early Modern England: Attending to the O-Factor* (Chicago: University of Chicago Press, 1999), 208.

10 Frederick Reynolds, *The Life and Times of Frederick Reynolds*, 2nd ed., 2 vols. (London: Henry Colburn, 1827), 1:110.

11 Steven Connor, *Beyond Words: Sobs, Hums, Stutters and Other Vocalizations* (London: Reakton Books, 2014), 40.

12 Connor, *Beyond Words*, 51–52.

13 On applause, see Baz Kershaw, 'Oh for Unruly Audiences! Or, Patterns of Participation in Twentieth-Century Theatre,' *Modern Drama* 42, no. 2 (2001): 133–54; Matthew Steggle, 'Notes Towards an Analysis of Early Modern Applause,' in *Shakespearean Sensations: Experiencing Literature in Early Modern England*, ed. Katharine A. Craik and Tanya Pollard (Cambridge: Cambridge University Press, 2013), 118–37. The catcall was an imitation of the yowl of a cat, produced by a whistle or by the voice.

14 Lamb, 'On the Custom of Hissing,' 123.

15 See Naya Tsentourou, 'Sighs and Groans: Attending to the Passions in Early Modern Prayer,' *Literature Compass* 12, no. 6 (2015): 262–73.

16 Anon, 'Police. Bow-Street,' *The Examiner*, no. 1246 (18 December 1831): 812.

17 E.P. Thompson, 'The Moral Economy of the English Crowd in the Eighteenth Century,' *Past and Present* 50, no. 1 (1971): 76–136. The literature on the English crowd and 'moral economy' is extensive: see, for example, George Rudé, *The Crowd in History 1730–1848: A Study of Popular Disturbances in France and England* (New York: Wiley, 1964); John Stevenson, *Popular Disturbances in England, 1700–1870* (London: Addison, Wesley Longman, 1979); Mark Harrison, *Crowds and History: Mass Phenomena in English Towns, 1790–1835* (Cambridge: Cambridge University Press, 1988); Nicholas Rogers, *Crowds, Culture, and Politics in Georgian Britain* (Oxford: Oxford University Press, 1998). None of these studies devote attention to tactics of crowd behaviour such as hissing. For a useful account and critique of Thompson's concept of the 'moral economy,' see Norbert Götz, '"Moral Economy": Its Conceptual History and Analytical Prospects,' *Journal of Global Ethics* 11, no. 2 (2015): 147–62.

18 Roger Wells, *Wretched Faces: Famine in Wartime England, 1793–1801* (Stroud: Alan Sutton, 1988), 128.

19 *Courier*, 15 September 1800.

20 *Oracle*, 16 September 1800.

21 Wells, *Wretched Faces*, 130.

22 *London Packet*, 15–17 September 1800; *Morning Post*, 16 September 1800; *Oracle*, 16 September 1800.
23 Wells, *Wretched Faces*, 130.
24 Quoted Wells, *Wretched Faces*, 130.
25 Thompson, 'Moral Economy,' 112.
26 John Marsh, *The John Marsh Journals: The Life of a Gentleman Composer* (1752–1828), ed. Brian Robins (Hillsdale, NY: The Pendragon Press, 1998), 194.
27 *True Briton*, 30 March 1795. See also Ian Woodfield, *Opera and Drama in Eighteenth-Century London: The King's Theatre, Garrick and the Business of Performance* (Cambridge: Cambridge University Press, 2001), 145.
28 Edmund Burke, *The Writings and Speeches of Edmund Burke*, vol. 8, *The French Revolution*, ed. L.G. Mitchell (Oxford: Oxford University Press, 1989), 8:119.
29 For an excellent discussion of the 1790s in these terms, see Georgina Green, *The Majesty of the People: Popular Sovereignty and the Role of the Writer in the 1790s* (Oxford: Oxford University Press, 2014).
30 E.P. Thompson, 'Patrician Society, Plebeian Culture,' *Journal of Social History* 7, no. 4 (1974): 402.
31 Weber, *Theatricality as Medium*, 34.
32 See Gillian Russell, *The Ephemeral Eighteenth-Century: Print, Sociability and the Cultures of Collecting* (Cambridge: Cambridge University Press, forthcoming).
33 For a more detailed discussion of these events, see Steve Poole, *The Politics of Regicide in England 1760–1850: Troublesome Subjects* (Manchester: Manchester University Press, 2000), 103–11, and John Barrell, *Imagining the King's Death: Figurative Treason, Fantasies of Regicide, 1793–1796* (Oxford: Oxford University Press, 2000), 556–59.
34 See *The Times*, 30 June 1795; Anon, *Account of the Proceedings at a General Meeting of the London Corresponding Society ... Held in an Enclosed Field, Behind the Long Room, Borough Road, St George's Fields, On Monday, the 29th of June, 1795* (London: D. I. Eaton, 1795). See also Mary Thale, ed., *Selections from the Papers of the London Corresponding Society 1792–1799* (Cambridge: Cambridge University Press, 1972), 252.
35 Anon, *Account of the Proceedings*, 10.
36 Anon, *Account of the Proceedings of a Meeting of the London Corresponding Society, Held in a Field near Copenhagen House, Monday, Oct. 26, 1795 ...* (London: Richard 'Citizen' Lee, 1795), 3.
37 Anon, *Account of the Proceedings of a Meeting of the London Corresponding Society*, 9; *Gentleman's Magazine* 65 (1795): 874.
38 Anon, *Account of the Proceedings of a Meeting of the London Corresponding Society*, 12.
39 Barrell, *Imagining the King's Death*, 553.
40 Anon, *A Narrative of the Insults offered to the King, On His Way To and From the House of Lords* (London: J. Owen, 1795); Anon, *Truth and Treason! Or a Narrative of the Royal Procession to the House of Peers, October the 29th, 1795* (London: no pub, 1795), 2.
41 Anon, *A Narrative of the Insults*, 13–14.
42 Anon, *Truth and Treason!*, 2; Anon, *A Narrative of the Insults*, 14.
43 Anon, *A Narrative of the Insults*, 14; Anon, *Truth and Treason!*, 2, 4.
44 Anon, *A Narrative of the Insults*, 15.
45 Anon, *Truth and Treason!*, 3.
46 Barrell, *Imagining the King's Death*, 556; Anon, *A Narrative of the Insults*, 17; Anon, *Truth and Treason!*, 3.

47 Anon, *Truth and Treason!*, 3.
48 Anon, *A Narrative of the Insults*, 18, 19.
49 A fragment of the coach window is preserved in the Museum of London: see Poole, *The Politics of Regicide*, 103.
50 Anon, *A Narrative of the Insults*, 19.
51 Anon, *A Narrative of the Insults*, 27.
52 Anon, *Truth and Treason!*, 7.
53 Anon, *A Narrative of the Insults*, 28.
54 Anon, *Truth and Treason!*, 7.
55 Anon, *Truth and Treason!*, 7.
56 Barrell, *Imagining the King's Death*, 559.
57 Quoted in Barrell, *Imagining the King's Death*, 574.
58 Anon, *Truth and Treason!*, 5; Anon, *A Narrative of the Insults*, 9.
59 Anon, *A Narrative of the Insults*, 9.
60 [Kidd Wake], *The Case of K. W.; Being a Narration of his Sufferings, During Five Years Confinement!!! in Glocester Penetentiary House: for Hooting, Hissing, and Calling out No War! as his Majesty was Passing in State to the House of Peers. ... Written by Himself* (London: Kidd Wake et al, [1801]), 6.
61 [Wake], *The Case of K. W*, 16–17.
62 See Marc Baer, *Theatre and Disorder in Late Georgian London* (Oxford: Clarendon Press, 1992).
63 'The Right of Hissing,' *The London Review of Politics, Society, Literature, Art and Science* (9 September 1865): 273.
64 For booing, see Dan Rebellato, 'Booing,' *Contemporary Theatre Review* 23, no. 1 (2013): 11–15.
65 Judith Butler, *Notes Towards a Performative Theory of Assembly* (Cambridge, MA: Harvard University Press, 2015), 8.
66 Walter T. Filbert, 'Prince Charles Attacked in Student Protests–London 9 December 2010,' *YouTube*, accessed 17 February 2016, www.youtube.com/watch?v=B7PcNpwwBmc.

Bibliography

Primary sources

Anon. *Account of the Proceedings at a General Meeting of the London Corresponding Society ... Held in an Enclosed Field, Behind the Long Room, Borough Road, St George's Fields, On Monday, the 29th of June, 1795*. London: D. I. Eaton, 1795.

Anon. *Account of the Proceedings of a Meeting of the London Corresponding Society, Held in a Field near Copenhagen House, Monday, Oct. 26, 1795*. London: Richard 'Citizen' Lee, 1795.

Anon. *A Narrative of the Insults Offered to the King, on His Way to and from the House of Lords*. London: J. Owen, 1795.

Anon. 'Police. Bow-Street.' *The Examiner*, no. 1246 (18 December 1831): 812.

Anon. 'The Right of Hissing.' *The London Review of Politics, Society, Literature, Art and Science* 11, no. 271 (1865): 273–4.

Anon. *Truth and Treason! Or a Narrative of the Royal Procession to the House of Peers, October the 29th, 1795*. London: no pub, 1795.

Congreve, Francis Aspry. *Authentic Memoirs of the Late Mr. Charles Macklin, Comedian*. London: J. Barker, 1798.

Courier.

Gentleman's Magazine.

Lamb, Charles. 'On the Custom of Hissing at the Theatres, with some Account of a Club of Damned Authors.' *The Reflector* 2 (1811): 122–27.

Lamb, Charles. *The Letters of Charles Lamb, with a Sketch of His Life by Thomas Noon Talfourd.* 2 vols. London: Edward Moxon, 1837.

London Packet.

Morning Herald.

Morning Post.

Plato, *The Laws.* Translated by Alfred Edward Taylor. In Plato, *The Collected Dialogues.* Edited by Edith Hamilton and Huntington Cairns. Princeton, NJ: Princeton University Press, 1973.

Reynolds, Frederick. *The Life and Times of Frederick Reynolds.* 2nd ed. 2 vols. London: Henry Colburn, 1827.

Times.

True Briton.

[Wake, Kidd], *The Case of K. W.; Being a Narration of His Sufferings, During Five Years Confinement!!! in Glocester Penetentiary House: for Hooting, Hissing, and Calling Out No War! As His Majesty was Passing in State to the House of Peers. ... Written by Himself.* London: Kidd Wake et al, [1801].

Secondary sources

Baer, Marc. *Theatre and Disorder in Late Georgian London.* Oxford: Clarendon Press, 1992.

Burke, Edmund. *The Writings and Speeches of Edmund Burke, Vol. 8, The French Revolution*, edited by Leslie G. Mitchell. Oxford: Oxford University Press, 1989.

Burke, Helen. 'Acting in the Periphery: The Irish Theatre.' In *The Cambridge Companion to British Theatre, 1730–1830*, edited by Jane Moody and Daniel O'Quinn, 219–31. Cambridge: Cambridge University Press, 2007.

Butler, Judith. *Notes Towards a Performative Theory of Assembly.* Cambridge, MA: Harvard University Press, 2015.

Connor, Steven. *Beyond Words: Sobs, Hums, Stutters and other Vocalizations.* London: Reakton Books, 2014.

Filbert, Walter T. 'Prince Charles Attacked in Student Protests—London 9 December 2010.' Accessed 17 February 2016. www.youtube.com/watch?v= B7Pc NpwwBmc

Götz, Norbert. '"Moral Economy": Its Conceptual History and Analytical Prospects.' *Journal of Global Ethics* 11, no. 2 (2015): 147–62.

Green, Georgina. *The Majesty of the People: Popular Sovereignty and the Role of the Writer in the 1790s.* Oxford: Oxford University Press, 2014.

Harrison, Mark. *Crowds and History: Mass Phenomena in English Towns, 1790–1835.* Cambridge: Cambridge University Press, 1988.

Horgan, Kate. *The Politics of Songs in Eighteenth-Century Britain, 1723–1795.* Abingdon: Pickering and Chatto, 2014.

Kershaw, Baz. 'Oh for Unruly Audiences! Or, Patterns of Participation in Twentieth-Century Theatre.' *Modern Drama* 42, no. 2 (2001): 133–54.

Marsh, John. *The John Marsh Journals: The Life of a Gentleman Composer (1752–1828).* Edited by Brian Robins. Hillsdale, NY: The Pendragon Press, 1998.

Poole, Steve. *The Politics of Regicide in England 1760–1850: Troublesome Subjects*. Manchester: Manchester University Press, 2000.

Rebellato, Dan. 'Booing.' *Contemporary Theatre Review* 23, no. 1 (2013): 11–15.

Rogers, Nicholas. *Crowds, Culture, and Politics in Georgian Britain*. Oxford: Oxford University Press, 1998.

Rudé, George. *The Crowd in History 1730–1848: A Study of Popular Disturbances in France and England*. New York: Wiley, 1964.

Russell, Gillian. *The Ephemeral Eighteenth-Century: Print, Sociability and the Cultures of Collecting*. Cambridge: Cambridge University Press, forthcoming.

Smith, Bruce R. *The Acoustic World of Early Modern England: Attending to the O-Factor*. Chicago: University of Chicago Press, 1999.

Steggle, Matthew. 'Notes Towards an Analysis of Early Modern Applause.' In *Shakespearean Sensations: Experiencing Literature in Early Modern England*, edited by Katharine A. Craik and Tanya Pollard, 118–37. Cambridge: Cambridge University Press, 2013.

Stevenson, John. *Popular Disturbances in England, 1700–1870*. London: Addison, Wesley Longman, 1979.

Tsentourou, Naya. 'Sighs and Groans: Attending to the Passions in Early Modern Prayer.' *Literature Compass* 12, no. 6 (2015): 262–73.

Thompson, E.P. 'The Moral Economy of the English Crowd in the Eighteenth Century.' *Past and Present* 50, no. 1 (1971): 76–136.

Thompson, E.P. 'Patrician Society, Plebeian Culture.' *Journal of Social History* 7, no. 4 (1974): 382–405.

Weber, Samuel. *Theatricality as Medium*. New York, NY: Fordham University Press, 2004.

Wells, Roger. *Wretched Faces: Famine in Wartime England, 1793–1801*. Stroud: Alan Sutton, 1988.

Woodfield, Ian. *Opera and Drama in Eighteenth-Century London: the King's Theatre, Garrick and the Business of Performance*. Cambridge: Cambridge University Press, 2001.

8 Laughed out of court

Counter-theatre and participatory justice in the trials of William Hone

Karen Crawley

On 18, 19 and 20 December 1817, William Hone, radical author and publisher, stood trial at the London Guildhall for seditious and blasphemous libel.[1] He had published three parodies that used familiar passages from the Church of England's *Book of Common Prayer* to attack the Tory government of Lord Liverpool.[2] Like many radicals tried for sedition during this repressive period of English history, Hone represented himself in court. The Attorney General argued that Hone's parodies had brought the sacred language of the religious texts into disrepute. Hone's defence appealed to the coercive force of precedent in English common law. Arming himself with books of liturgical parodies written by highly respected persons (including Martin Luther as well as Tory politician George Canning, a target of Hone's parodies), he read these parodies aloud, one after the other, thereby demonstrating that the liturgical form could be, and frequently was, a vehicle for satire, rather than its object.[3] As Olivia Smith has noted, 'Hone was probably the only radical in history who could legitimately defend himself by claiming that he had attacked the state.'[4] Hone's virtuosic performance, in which he spoke for over 22 hours over three days, took place in front of a responsive audience, who laughed at the recitation of his parodies, cheered him when he scored a point, coughed at the absurdity of the Attorney General's rebuttals and guffawed at the pompous and ineffectual attempts of the presiding judge (who was, on the second and third day, Lord Ellenborough himself, also a target of Hone's parodies) to keep order in the court. At the conclusion of each day, the jury declared Hone 'not guilty,' to the delighted applause and cheering of the 1,000 people in the packed courtroom, and the crowd of 20,000 waiting outside. In showing that his parodies were political rather than blasphemous, Hone exposed the libel charges as a pretext for censorship, and emerged as a kind of political martyr, a common man who had bravely stood up to the forces of repression which, as Hone later wrote, had been 'laughed out of court.'[5]

Blasphemous and seditious libel prosecutions were a favoured tool used by the government to repress radical behaviour and free thought in the late eighteenth and early nineteenth century. The period of particularly vehement state repression in the years after Waterloo in 1815 was also 'the heroic age of libel trials,'[6] when a number of key radicals such as William Cobbett, Thomas Wooler and Richard Carlile were charged with libel. In 1817, the year of Hone's trials, there were more than 20 prosecutions of radical pressmen for seditious libel, and many more were imprisoned without trial.[7] These trials pitted the Tory government against radical or progressive publishers and writers who not only celebrated free thinking, but also enacted it. The odds were stacked against the accused. Throughout this period, juries were stacked,[8] high court judges held office through government appointment and prosecution counsel were experienced – the Attorney General often acting on behalf of the Crown. Nevertheless, the requirement that criminal law matters be dealt with in the form of a public trial forced radicals and government into what James Epstein calls 'a rather curious dialogue.'[9] Radicals had to answer to charges, but they also had the right to have their defence heard, and often chose to defend themselves.[10] As Marcus Wood has noted, libel trials were central to the radical movement, providing an officially sanctioned context in which the state had to justify itself under public scrutiny.[11] They helped to sustain the links between the printed and spoken word after the radicals had their political and public group presence limited by measures like the *Seditious Meetings Act*, passed in March 1817. Libel trials brought radicals and government officials/representatives together in an intertwined and unscripted theatre, a dramatic, unstable encounter. In the case of Hone, whom Epstein calls 'the greatest radical satirist of the Regency era,'[12] the outcome was decisive. A contemporary ministerial paper, referred to in James Routledge's 1876 history of popular radicalism, asked whether 'a crowded court laughing at and sneering against the guardians of the law' should be considered 'a victory.'[13] After Hone's acquittal, the government did not file a prosecution for libel for almost two years, and by the 1820s, they had all but abandoned libel for being an ineffective mechanism of control.[14] A large part of the reason why, as Hone's case highlights, is that the English courts in this period echoed a barrage of radical and popular dissent.

English criminal justice depends on an audience. Publicity, wrote Jeremy Bentham, is 'the very soul of justice.'[15] The integrity of criminal proceedings is bound up in their being held in public places in which spectators unconnected with the trial are able to observe justice being done.[16] Crucially, a criminal trial must have an audience not only for them to observe that justice is being carried out, but to be co-opted into participating in a performance of legal authority. As scholars of legal history and legal aesthetics have long recognised, the staging of trials – their language, rituals, costumes and architecture – is intended to enact the law's authority and legitimise the exercise of state power.[17] The importance of ritual, symbolism and theatre within the English criminal justice system is well known. The law's constitutive dependence on such theatricality is a double-edged sword: in order to preserve its own claims to impartiality and rationality, the law must suppress its own tendency towards spectacular, emotional, visual theatricality.[18] The control the judge exercises over sound in the court, over who speaks and when, forms an inextricable part of the state theatre that conveys the majesty of law, and in which 'silence is understood as a condition and a sign of the proper administration of justice.'[19] An atmosphere of formality and solemnity is crucial to the state's performance of its 'legitimised symbolic violence.'[20]

It is well known that eighteenth century trials often fell short of this ideal. They were frequently conducted in an atmosphere of noise and disorder.[21] That noise came from members of the public who were gathered to witness the proceedings, a public that used to be portrayed as a rabble or group or crowd but were increasingly becoming a literate public sphere. Women and men of all classes continued to attend court in equal numbers until well into the nineteenth century. They regarded the court as a place of entertainment to which they flooded when there were sensational cases or particular barristers were appearing. Peter King has written of the 'participatory theatre' in courtrooms in this period, in which the performance of judicial authority was balanced by the 'counter-theatre' of the crowd, who cheered and catcalled, booing and hissing at witness testimony, while applauding their favourites.[22] Such interruptions by a lively audience were a common feature of trials, and of trial reporting in newspapers and published pamphlets. Court reporters were always in attendance, and a transcript of the proceedings in Hone's case was published each day in *The Times,* and by Hone himself the following year.[23] They were described by E.P. Thompson in *The Making of the English Working Class* as 'some of the most hilarious legal proceedings on record.'[24] To a reader of modern trial literature, their most arresting feature is that they are frequently marked by square brackets capturing the sounds of a responsive audience: cheering, laughing, applause, coughing, huzzaing and making 'expressions of feeling.'

This chapter rereads the transcripts of Hone's trials that appeared in the newspaper and were edited by him, focussing on those moments of interruption where the solemnity of the Court was at stake: the audience's laughter upon first reading of the charges, the humiliation of Lord Ellenborough, and the thunderous clapping and shouts of jubilation that occurred at the conclusion of each trial. An increased attention to how these sounds worked in the trial space, and in the trial transcript, allows us to understand the government's defeat in Hone's case as a result of the embodied and aesthetic experience of an audience participating in law. It was Thompson, of course, who elsewhere coined the phrase 'counter-theatre,'[25] and its bipartite precision is useful here. In the state's theatre, the actors perform with full agency and didactic presence; the audience's role is to be passive: to watch, listen and learn. In the law's counter-theatre, the public gallery finds a way to participate, not just in the margins of the court's sovereign rituals, but by disrupting them entirely. The English legal system's openness to audience meant that the state's theatrical performance must be powerful to be seen as legitimate, but it also opened the potential of a counter-theatre that would disrupt the state's power over law. The audience's interruptions transformed Hone's trial from a univocal theatre of state power to a counter-theatre that signalled the triumph of law over politics. It took place against the broader shift in which the unruly and undifferentiated crowd gradually became a literate public to whom the law was accountable.

Hone's defence

Hone was tried at the London Guildhall, the hub of the city where all aspects of local administration were conducted, suggesting the government's confidence that this was the occasion when its powers would be demonstrated in front of the whole nation. Having previously worked as a legal clerk, Hone had read and published trial literature, and knew that trials were 'exciting vehicles of political education.'[26] The second trial attracted thousands of Londoners who began to queue from the early hours of the morning, blocking the roads and lanes surrounding the court. When the doors opened at nine, a thousand individuals managed to cram into the Great Hall, leaving 20,000 outside to await news of the day's proceeding.[27] The audiences for this kind of performance were, of course, not limited to the jury or spectators present at the trial, but to a wider, newspaper-reading public with an appetite for legal drama. Trial texts were a popular form of political literature, published in newspapers and radical journals, as well as being sold separately in cheap pamphlet form. Hone was personally acquainted with many of the prominent radicals of the day, and in preparation for his trial, Hone drew on a network of associates who allowed him access to their libraries and helped assemble his mountain of references, and even rehearsed his presentation in George Cruikshank's studio.[28] Smith notes that *The Trial of John Lilburne* (for publishing pamphlets criticising Cromwell in 1649) 'furnished a map of strategy and resistance'[29] for Hone, who presented himself initially as a solitary, honest man persecuted by a tyrannous and corrupt government and then, increasingly, in the persona of a martyr.[30] Depicting himself as a good upstanding Christian, he insisted that he was not a blasphemer, but the critic of a corrupt government, who had been labelled a blasphemer because ministers thought this was the easiest way to secure a guilty verdict from a jury that was likely to take far greater offence at slurs against church doctrine than slurs against the government. Hone insisted that his parodies were not blasphemies but political squibs, and that if the jury found them to be political squibs, they must acquit him. 'There were two kinds of parodies,' he explained to the jury,

> one in which a man might convey ludicrous or ridiculous ideas relative to some other subject; the other, where it was meant to ridicule the thing parodied. The latter was not the case here, and therefore he had not brought religion into contempt.[31]

Hone freely admitted to having ridiculed the government, but this had not been the meaning that the government had ascribed to his works, and he should therefore be found not guilty.

The heart of the case, the law of libel, was notoriously slippery. Libel was defined as 'any form of printed matter, the content of which had a tendency to provoke a breach of the peace.' On each of the three days his trials lasted, Hone began by describing the law of libel, which he called 'a shadow,' an 'undefinable crime.'[32] Hone's strategy effectively hollowed out the legal concept of libel, making it clear how it could readily be used as a tool for repression. Since *Fox's Libel Bill* of 1792 made the jury judge of the matter as well as the fact, Hone offered a counter-definition: 'That only could be called a libel which twelve men, sworn well and truly to try the cause, declared to be one.'[33] Hone thus casts the jury and defence as representatives of a vulnerable and honest state, and sets himself 'as a man of plain understanding'[34] against the tricky language and ceremony of his persecutors. His defence consisted in reading out from the rich history of liturgical parodies he had collected for a history of satire he intended to write. Hone's defence appealed to the strongly coercive nature of the English doctrine of precedent, with its unbreakable chain to past cases.[35] His performance both relied on and attested to the existence of a common literary heritage which had been and could be used in parody without belittling the model.[36] Much rested on the fact that Hone, as Clara Tuite writes, was

> an able performer of his own written texts. His self-defence became a virtuosic, embodied performance of the printed parodies, fully realizing Hone's conception of the printed text as a performance event and enlisting the public's support through their involuntary laughter and applause.[37]

In a moment of rhetorical triumph during the first trial, Hone signalled that George Canning, the Cabinet Minister who was one of his persecutors, had also published a liturgical parody in the *Anti-Jacobin* newspaper. 'He (Hone) was dragged before the Court, from behind his counter – and for what? For doing that which a Cabinet-Minister had been suffered to do with impunity.'[38] Hone waved the newspaper in front of the court, pointing out that Canning had even had the parody illustrated by Gillray, who, Hone pointed out, 'had a pension for his parodies.'[39] The *Times* reported: 'Here there was great applause from the Crowd assembled at the extremity of the Court.'[40] Similarly, Hone's account revealed: 'a number of persons in Court here applauded.'[41] Justice

Abbot declared that 'if such indecent interruptions were persisted in, he would order the Court to be cleared.'[42] *The Times* reported:

> Mr Justice Abbot expressed his displeasure at this interruption of the Court, and stated, that if any person should be pointed out as guilty of it, he would take care that such person should not again have an opportunity of disturbing the Court on that day.[43]

In a section omitted from *The Times* coverage, the trial transcript records that 'Mr. Hone hoped that the Attorney-General would bring Mr. Canning to justice (*Cheering*).'[44] Here the cheering of approbation served as a kind of public bluff – a dare to place his governmental persecutors in court to test the meaning of the law.

In order to keep within the bounds of his argument, Attorney General Shepherd was forced to acknowledge that all the parodies that Hone had mentioned were in fact libels, which meant that all those eminent men, including Luther, had also been blasphemers, 'which, no doubt, in their later lives they often repented, more especially when the time arrived that they were to settle their account between their own consciences and their God.'[45] The transcript notes in square brackets: '[Violent coughing, and other marks of disapprobation, on the part of the spectators, here interrupted the Attorney-General],'[46] which prompted another round of threats to clear the courtroom by Mr Abbot if order was not observed. Making no effort to contradict Hone's argument that his was a political and not a religious parody, Shepherd only managed to convey the impression that he was 'casting aspersions upon the soul of the father of Protestantism.' The jury deliberated for less than 15 minutes.[47] Upon the reading of the verdict, *The Times* reported the sound of the audience as well as the response to that sound by the judge:

> ... acclamations of applause were instantly heard from all parts of the Court; *Long live an honest jury, and an honest jury forever*, were exclaimed by many voices: the waving of hats, handkerchiefs and applause continued for several minutes. When order had been somewhat restored, Mr. Justice Abbott interposed, and desired that those who felt inclined to rejoice at the decision, would reserve the expressions of their satisfaction for a fitter place and opportunity.

The report in *The Times* ended here, but Hone's version added the following sentence:

> The people accordingly left the Court, and as they proceeded along the streets, the language of joy was most loudly and unequivocally expressed; every one with whom they met, and to whom they communicated the event, being forward to swell the peal.[48]

Indecorous laughter

Hone's parodies, as well as the trial transcripts which reproduced them, were circulated and celebrated within a culture of oral performance, and disseminated through an informal but efficient, nationwide but decentralised, and highly responsive radical publishing network that mixed together oral and written cultures.[49] Hone's publishing career showed his expertise in using oral texts – nursery rhymes, Christmas carols and English church liturgy – as the basis of his parodic technique. In taking the formal shell of a familiar cultural text with a ritualistic form, he could reach semi-literate members of the working classes who could be expected to know a number of the texts he was parodying from British oral tradition. Each of Hone's prosecuted parodies was liturgical, drawing on a genre inscribed and reinscribed through verbal and communal repetition.[50] The parody at issue in the first trial, *The Late John Wilkes's Catechism,* imitated the catechism, the manual of questions and answers memorised by believers and uttered both in private and public. Catechism, to *sound down*, evokes a sonorous ritual, an embodied experience of reciting in unison; of hearing oneself speak in unison with others. As Walter J. Ong has observed, 'sound forms community as reading alone cannot.'[51] By the early nineteenth century, catechisms had become a widely used form of popular entertainment.[52] The familiarity of the form made parodies widely accessible. Their oral quality facilitated communal reading practices in social spaces such as coffee houses, alehouses, street corners and reformist meetings.[53] Hone's parodies were read aloud and performed at taverns, along with other venues of varying respectability, and they were enormously popular throughout England in 1817 even before the December trials. They formed part of the alehouse counter-theatre of the Spenceans, a group of utopian, proto-socialist radicals who met to engage in debates and satirical political toasts, and who were one of the targets of the *Seditious Meetings Act* of 1817. The questions in Hone's *Political Litany* would have been read out by patrons, while the responses would have elicited a chorus of voices.[54] Kyle Grimes argues that by collapsing traditional 'speaker/writer, listener/reader relations,' Hone's parodies formed part of a broader shift in British popular readership towards what he calls 'the birth of the body politic':

> ... a shift from an audience constituted as an aggregate of the relatively passive and isolated 'receivers' of a text that emanates from some central author or authority, to an audience constituted as a set of persons who themselves enact and embody a virtually authorless and freely circulating text and whose ideological unity is signalled in their reflexive, immediate and contagious laughter.[55]

Historians of radical politics often understand laughter as inherently subversive, calling it 'profoundly seditious,' embodying an 'immediate, collective and infectious anti-authoritarianism.'[56] They also emphasise how laughter generates community – 'the unifying social bond of laughter' works to cement 'the bonds of a common ideology.'[57] For Mikhail Bakhtin, the laughter of the folk people was a regenerative and liberating noise, a dramatic bodily effusion that challenged the repressive seriousness of official authority.[58] The laughter of those at the bottom of a hierarchy has a kind of power, a 'levelling force,'[59] however temporary or insurgent, that overturns hierarchical distinctions and undermines those in positions of power. When, at the beginning of Hone's trial, Attorney General Shepherd read out the charge, quoting from Hone's parody of the scripture: 'I, the Minister, am the Lord thy liege, who brought thee out of want and beggary into the House of Commons,'[60] the sound of the government's legal representative reciting Hone's phrases of ministerial omnipotence was too much for some members of the crowd, who broke into laughter. The *Times* transcript marks in square brackets: 'Here an expression of feeling was manifested by some individuals of the Crowd in the Hall of the Court.' The audience's laughter here suggests that they have moved from theatre, in which their attention is fixed by the seriousness of courtroom procedure (which in this case allows, nay *necessitates*, a public reading of proscribed texts) over to the radical public counter-theatre, in which the audience participates in the public performance event of Hone's parodies in line with their original intentions. Put simply, the court was losing control.[61]

Justice Abbot halted proceedings and explained to the assembled crowd that the successful functioning of legal proceedings turned on an atmosphere of solemnity and gravity:

> If there is anybody present of so light a disposition as to think that a matter of this kind should be made a subject of laughter, at least he shall learn that he shall not come here to interrupt those who are of a graver disposition, and in the discharge of an important duty.

The Attorney General matched his humourless pomposity:

> My lord, if there be any persons here who can raise a smile at the reading of the Defendant's publication, it is the fullest proof of the baneful effect it has had, and with which I charge it. It is for this very reason I charge it as a libel on the Law of England. I am not sorry for the faint smile just uttered in Court. It establishes the baneful tendency of the work. If there be any here who are not Christians of some sect or another, God forbid that I should have their applauding support.[62]

The Attorney General steps in to retrieve control by claiming the laughter under the heading of evidence, which would make an interruption that is apparently favourable to the accused a part of the case against him. He tried again the second day when the audience once again laughed: 'If the social bonds of society are to be burst asunder by the indecent conduct of a rabble, the Court may as well discontinue its proceedings.'[63] This blunt contrast between the intelligence and good sense of the jurors with the base passions of the mob did not endear him to those present. His words draw on the idea of laughter as an irresistible contagion, an unruly force, a manifestation of the insouciance of the lower classes, recalling Lord Chesterfield, in his famously pompous eighteenth-century letters, who wrote 'there is nothing so illiberal, and so ill-bred, as audible laughter' and 'loud laughter is the mirth of the mob, who are only pleased with silly things.'[64] His comment is an obvious attempt to hide the fact that the spectators' laughter here is not inimical to their own morality, but corrosive of the law's authority.

As historians of laughter have noted, laughter became an increasing focus of regulation over the nineteenth century and became a matter of individual comportment precisely because of its collective and social power. Humour itself was undergoing a transformation as coarse, bawdy, scatological and obscene mirth was being reshaped into reasonable, decorous and polite comedy. Hone's trial occurred at the end of what Vic Gatrell deems the golden age of satire, right before the newly refined 1820s.[65] His analysis of the comic prints of this period (comic spectacles, not moral lessons) in which artists like Cruikshank took aim at the excesses of the rich, the corruption of the political elite and the royal family shows a society where the well born mingled with the lower orders, laughing at rather than condemning drunkenness, lechery and idle pursuits. Accordingly, Gatrell observes that 'a high-mindedness co-existed with low-mindedness most nakedly in London.'[66] The growing cultural importance of 'politeness' over the eighteenth century meant that by the beginning of the nineteenth century, manners had been largely domesticated and humour tamed, as an increasingly influential middle class defined itself in opposition to both the labouring classes and the aristocracy. Hone's humour and speaking style perfectly navigated these changing waters: he emphasised that his squibs were political, not bawdy; he presented himself as an upright Christian.

Here, the cleverness of Hone's strategy revealed itself at those moments where he aligned himself with the court's own formal procedure, distancing himself from the laughter in the courtroom and emphasising the solemn and serious nature of the proceedings. After Justice Abbot's defeat, the Chief Justice presided on the second day, determined to convict Hone.[67] A much-despised judge, Ellenborough had a fierce reputation for terrorising defendants.[68] He immediately announced his priority to stop the crowd laughing and cheering as it had done the previous day, sending for the sheriffs so that he could enforce order in the court. The Attorney General read out the sacred words of the service book:

> Son of God, we beseech thee to hear us!
> O Lamb of God, that takest away the sins of the world, have mercy on us.[69]

This was then followed by Hone's parody of those words:

> Son of George, we beseech thee to hear us.
> O House of Lords, that takest away so many thousands of pounds in pension, have mercy on us.[70]

Once again, members of the audience found the Attorney General incanting these mock-solemn words very funny. *The Times* notes an eruption of 'indecorous laughter';[71] Hone's *Trials* add an editorial comment:

> These parts of the parody produced an involuntary burst of laughter from the auditory, which evidently proceeded, not from a wish to disturb the Court, but was really the irresistible impulse arising from the matter of the parody.[72]

Hone's editorial comment reframes the spectator's laughter not as a breach of the peace but as a natural effect of the parody functioning as it should; it absolves the spectators of fault or 'indecorousness.' Hone's response to the audience's laughter, both during the trial and in his published *Trials,* was complex and clever. He immediately jumped up and attempted to address the issue of order but was shut down by Lord Ellenborough as speaking out of turn. Hone replied: 'I was only desirous, my Lord to add my feeble assistance in keeping order in the Court,' to which Lord Ellenborough replied 'A very proper disposition.'[73] At this point the Sheriff arrived, and Lord Ellenborough explained that he had sent for him to maintain order, as 'there was a most unseemly disturbance in the Court yesterday, I understand, and there has been another today.'[74] When Hone was allowed to resume speaking, he announced that prior to addressing the Attorney General's charges, he would address the persons present. The transcript records Hone's words in third person:

> ... and he must say, he knew of no species of indecency – he knew of no breach of propriety, that more deserved severe reprehension and reprobation, than an act which tended to impede the proceedings of a Court of Justice. Taking his trial there, on a charge which perhaps might consign him to a prison, he felt most solicitous that good order and tranquility should prevail. Perhaps those indiscreet expressions of feeling might increase his danger – they certainly could not

> serve any good purpose. The persons who had so loudly expressed themselves ought not, whatever they might think or feel, to have given loose to their feelings in that place. He begged to state, that he was opposed to every such expression in that Court; and he declared that man to be his decided enemy – he cared not where he came from, or who he was who should attempt to interrupt such solemn proceedings.[75]

As an experienced and bold radical satirist, Hone was a master in the theatre of spoken word, direct contact and collective feeling, elements of popular life that were increasingly organised in this period as a form of political experience.[76] He disavowed the disruption to court proceedings by appealing to the dignity of the law and its solemnity, which he then expertly detached from the corrupt officials in court, and resituated in his own relationship with his audience as members of a new legal public. The transcript published by Hone added the following editorial comment at the end of his defence speech on the first day:

> Mr Hone, towards the end of his speech, was much affected. In the course of it there were some tokens of applause in the Court, but they were soon silenced by those who felt the impropriety of such demonstrations. He was also interrupted by Mr Justice Abbot in some other instances not noticed, but the substance of the learned judge's objections to the course adopted by Mr Hone in his defence has been stated.[77]

In this passage, Hone is clearly writing as the protagonist; it is his speech that is 'interrupted' by Justice Abbot, rather than the audience 'interrupting' the court proceedings. In contrast to Shepherd's undifferentiated notion of the 'rabble,' the crowd always threatening to erupt in inappropriate ways, Hone affirms that 'such demonstrations' were made by a limited number of those present in the audience, and that the majority of the Crowd respected the formality of the court environment.

Justice Ellenborough fared no better against Hone than Justice Abbot. Although Ellenborough insisted that Hone's citation of parody was inadmissible – 'No person, under pretence of explaining one libel, shall offend the ears of public decency by the recital of such profanations'[78] – Hone simply got on with it, stating: 'I must go on with these parodies, or I cannot go on with my defence.'[79] Soon after, Hone goaded Lord Ellenborough into making a fatal mistake, when he pronounced Hone's parodies libel 'in law and fact.'[80] This, Hone pointed out triumphantly, meant he had judged the question that it was the jury's role to decide. Hone declared in what the transcript calls 'a triumphant tone': 'His Lordship has declared his opinion, but let me say that, after all, it is the opinion of *one* man, it is but his Lordship's opinion. Of course I speak this in no offensive sense.'[81] The transcript records 'loud huzzaing.' *The Times* reports 'some disturbance.'[82] Ellenborough replied:

> So I understand, but it might be as well if a little decency were preserved at the bottom of the Court. If the officers take any person into custody who makes a disturbance, let him be brought up to me, and I will reward such conduct.

The sheriff leapt up and told 1,000 laughing men and women: 'The first man I see laugh, after such a severe notice, shall be brought up.'[83] The crowd again cheered Hone's response to one of Ellenborough's interruptions, when Hone cried 'it is I who am upon my trial, not your Lordship!'[84] and Ellenborough directed one of the sheriffs to leave the bench and go to the bottom of the court to quell the disturbance. 'It is impossible that the officers can be doing their duty; let them bring any man before me, and I will put an end to this.' But the sheriffs could not pick out one laughing person among a thousand, so the officer of the court had to report that no one had been caught. Ellenborough responded: 'Open your eyes, and see; and stretch out your hands, and seize. You must have observed somebody. Mark where the noise comes from, and note the man.'[85]

This was the most disruptive moment of Hone's trials, in which the audience, expected to be passive and silent, erupted on a collective scale. In the figure of the judge, angry and humiliated in his own court, and the court officials who can do nothing in the face of the collective power of laughter, we see the dignity of the courtroom is reduced to dependence on the cooperation of the assembled crowd. This hilarious spectacle calls to mind the older plebeian tradition of folk theatre, or counter-theatre in the sense in which Thompson first used it: many ballad singers, street-corner parodies and satirical monologues of this period, as well as much of the entertainment in taverns and coffee houses, took the form of counter-theatre.[86] Characterised by a mocking or lampooning of the texts, offices and ceremonies of the establishment, it undermined traditional authority through play and reversal and did so in a way that could be understood across working-class and upper-class, literate and non-literate, lines. Ellenborough never quite recovered from this moment of counter-theatre which so clearly exposed his tenuous legal authority. At the end of the second day, both *The Times* transcript and Hone's publication reports that Ellenborough 'appeared much oppressed with indisposition during the latter part of the trial' and 'delivered his charge in so faint a tone, that it was scarcely audible beyond the Bench.'[87]

The people's law

After Hone was acquitted for a second time, the morning edition of *The Times* confidently predicted that Hone's third trial would be cancelled.[88]

The tension on the third day is evidenced by Hone's introduction, which said 'Though the Court was crowded to excess, the most profound silence prevailed.'[89] The crowd in attendance as well as the broader newspaper reading public was most definitely with Hone; Epstein suggests that the failure of many of the carefully vetted special jurors to show up at Hone's second and third trials may have been the result of juror's fears of the London populace.[90] Having spoken for six hours on the first day, and seven on the second, Hone was, at the beginning of the third trial, 'extremely agitated' and 'seriously indisposed.'[91] The Attorney General acknowledged that 'he appeared to be very unwell,' but Hone did not request a delay.[92] After some time spent gathering himself together while the charges were read out, Hone came out swinging against Lord Ellenborough, arguing, 'with increased vehemence,' that he had 'sustained an injury' from the judge, who had offered his opinion in contradiction to the Libel Bill.[93] Speaking 'earnestly and slowly,'[94] (the transcript records), Hone said 'I think it necessary to make a stand here,' and explained that Lord Ellenborough had interrupted him many times. Turning towards the jury, Hone emphasised their absolute power and the judge's secondary position as legal adviser to the court – 'His lordship is no judge of me. You are my judges, and you only are my judges. His lordship sits there to receive your verdict.'[95] Indeed, Hone explained, the judge 'is not even there to regulate the trial – for the law has already regulated it. He sits there only as an administrator of that law.'[96] By making this point, Hone separated the biased judge from the impartial law, which was on his side. Hone said he hoped that today Ellenborough would give his opinion 'coolly and dispassionately, without using either expression or gesture which could be construed as conveying an entreaty to the jury to think as he did,'[97] thus masterfully putting Ellenborough's facial and bodily movement under scrutiny from the audience. He concluded this preface by appealing to the audience to remain calm also:

> He should regret much, if in the course of his trial any such expression of feeling should be manifested as occurred yesterday. They who were present ought to command themselves, and remember that he was on a trial of life or death.[98]

(He was not facing a death sentence, as he went on to explain, he meant that a guilty verdict would deprive him of health and life.)

In this third and final trial, Hone's performance and the crowd's responsive sounds reached a new intensity. Throughout this speech on his final day, more expressions of support from the crowd are noted in rounded brackets: 'Murmurs of Applause,'[99] 'Murmurs from the Crowd.'[100] Hone frequently referred to the jury not only as his judges, but also as his equals and friends and, as Wood notes, constantly alludes to the acts of standing and sitting during trials in order to emphasise their discomfort at being assembled thus.[101] In all three trials, he gives a detailed account of his arrest under ex officio information and his subsequent mistreatment during the months leading up to his trial,[102] but in this third trial, he leveraged his appearance of bodily exhaustion into his performance, claiming that 'Providence' was intervening to help 'a defenceless man against the rage and malice of his enemies,' and that 'he was weak and incapable, and was at that moment a wonder unto himself.' The transcript notes in rounded brackets: 'Here a mixed murmur of applause and pity was heard from the crowd assembled.'[103] Hone dramatises the physical violence of the court's persecution and is cheered on: 'He can't stand the third trial' said these humane and Christian ministers, 'we shall have him now, he must be crushed. (Great shouts of applause.)'[104] After reciting once again a litany of parodies written by eminent men, the transcript records: 'Who, O who (cried Mr Hone, raising his voice to a tone of the utmost vehemence), who were the blasphemers?' 'Great applause instantaneously burst from every part of Guildhall; and Lord Ellenborough declared he would adjourn the Court if greater order was not observed.'[105] The jury deliberated for 20 minutes before returning.

> The Court was exceedingly crowded; and as soon as the decision was heard, loud and reiterated shouts of applause ensued. His lordship called upon the sheriffs to preserve order; but the expression of feeling was so universal, that all interposition was impossible. The crowd then left the Court, and, mixing with the multitude in the hall and in the passages, communicated their feelings with their news.

The Times simply concludes: 'the shouts were repeated.' Hone's *Trials* add:

> the loudest acclamations of applause filled the avenues, and were echoed through Guildhall and King Street, which were extremely crowded. Never, indeed, was a greater degree of public interest excited upon any trial. The Court was crowded throughout the day; and for several hours before the jury retired, Guildhall was as full as upon the assemblage of a Common Hall. The sensation produced by the result of this important trial cannot be described.[106]

The Hone trials vividly highlight the tension between the judge's attempts to constitute legal authority as the domain of dispassionate, controlled and quiet reflective judgement, and the radical noise of Hone and the spectators who transformed the court into a counter-theatre. This final description of the assembled crowd, moving and interacting with one another in celebrating Hone's acquittal, locates that counter-theatre as a moment of spontaneous and shared joy; the joy of an embodied, physical experience of participatory justice. The openness to spectators is a structural requirement of theatre *and* justice. As Linda Mulcahy has recently argued, members of the public who are admitted as spectators to trials are not just passive auditors there to be educated or to provide a check on the impartiality of the judge. Audience members evaluate, critique and object to procedure and decisions. Their own agentic role in the process is a crucial part not just of open justice but of participatory justice – they are participants rather than spectators – and in this sense they allow for a collective experience of law. For Mulcahy, who charts the ways in which the public has been sidelined in modern courtrooms, the trial's moral legitimacy rests on efforts to include the public in the justice system as participants rather than mere spectators.[107] The legitimacy of a trial relies on its openness, but that openness is what allows for a counter-theatre of justice, where spectators do not merely observe and ratify justice, but participate in it by calling the forms of legal authority to account. By listening to Hone's transcript, we can hear the emergence of the modern public of law in cheers and murmurs and peels of raucous laughter.

Notes

1 In the first two trials, Hone was charged with blasphemous and seditious libel, and in the third simply with blasphemous libel. The arguments all hinged on the latter: see Marcus Wood, *Radical Satire and Print Culture 1790–1822* (Oxford: Clarendon Press, 1994), 100.

2 Published in January 1817, the three parodies included: *The Late John Wilkes's Catechism of a Ministerial Member*, *The Political Litany, Diligently Revised* and *The Sinecurist's Creed, or Belief.*

3 'It is an important feature of my defence, to show that parodies might be written, in order to excite certain ideas, without any desire to turn the original production into ridicule.' William Hone, *The Three Trials of William Hone, for Publishing Three Parodies; viz. The Late John Wilkes's Catechism, The Political Litany, and the Sinecurist's Creed; on Three Ex-Officio Informations, at Guildhall, London, on Three Successive Days* (London: Printed by and for William Hone, 67, Old Bailey, 1818), 55.

4 Olivia Smith, *The Politics of Language, 1791–1819* (Oxford: Clarendon Press, 1984), 180.

5 William Hone, 'Address' (23 January 1818) in Hone, *Three Trials,* i.

6 Clara Tuite, 'Not Guilty: Negative Capability and the Trials of William Hone' in Nicola Moore, *Censorship and the Limits of the Literary: A Global View* (London: Bloomsbury Publishing, 2012), 34.

7 *Suspension of Habeas Corpus Act* (1817).

8 The furore over the apparently corrupt jury-selection process (whereby a small number of handpicked 'special jurors' would tilt the jury in favour of the government) in the trial of Hone's fellow radical T.J. Wooler in June 1817 led to the reform of the City of London special-jury list, and the untainted jury pool probably helped Hone's acquittal a few months later. See Philip Harling, 'The Law of Libel and the Limits of Oppression, 1790 to 1832' *Historical Journal* 44, no. 1 (2001): 117.

9 James Epstein, *Radical Expression, Political Language, Ritual and Symbol in England, 1790–1850* (Oxford: Oxford University Press 1994), 33.
10 Epstein, *Radical Expression,* 32.
11 Wood, *Radical Satire,* 98. See also Michael T Davis, 'Prosecution and Radical Discourse During the 1790s: The Case of the Scottish Sedition Trials,' *International Journal of the Sociology of Law* 33, no. 3 (2005): 148.
12 Epstein, *Radical Expression,* 35.
13 James Routledge, *The History of Popular Progress Chiefly in Relation to the Freedom Of and Trial By Jury 1660–1820* (London: Macmillan, 1876), 438.
14 Wood, *Radical Satire,* 98, Kevin Gilmartin, *Print Politics: The Press and Radical Opposition in Nineteenth-Century England* (Cambridge: Cambridge University Press, 1996), 123. Their focus shifted to unlawful assembly: see Harling, 'The Law of Libel.'
15 'Draught of a Code for the Organisation of the Judicial Establishment in France' (1790), in *The Works of Jeremy Bentham*, ed. John Bowring, 11 vols. (Edinburgh: William Tait, 1838–43), 4:316 cited in David Lemmings, 'Introduction: Criminal Courts, Lawyers and the Public Sphere,' in David Lemmings, ed. *Crime, Courtrooms and the Public Sphere in Britain, 1700–1850* (Farnham: Ashgate, 2012), 1.
16 Linda Mulcahy, *Legal Architecture: Justice, Due Process and the Place of Law* (London: Routledge, 2011), 85.
17 See, for instance, the work of Peter Goodrich, *Languages of Law: From Logics of Memory to Nomadic Masks* (London: Weidenfeld and Nicolson, 1990); 'Specula Laws: Image, Aesthetic and Common Law,' *Law and Critique* 2, no. 2 1992): 233; 'The Continuance of the Antirrhetic,' *Cardozo Studies in Law and Literature* 4, no. 2 (1992): 207. See also Clare Graham, *Law: The Architectural and Social History of the English Law Court to 1914* (London: Routledge, 2003).
18 Julie Stone Peters, 'Legal Performance Good and Bad,' *Journal of Law, Culture and the Humanities* 4, no. 2 (2008): 181–82.
19 James Parker, *Acoustic Jurisprudence: Listening to the Trial of Simon Bikindi* (Oxford: Oxford University Press, 2015), 3.
20 Pierre Bourdieu, 'The Force of Law: Toward a Sociology of the Juridical Field,' *The Hastings Law Journal* 38, no. 5 (1987): 838.
21 Amy Milka and David Lemmings, 'Narratives of Feeling and Majesty: Mediated Emotions in the Eighteenth Century Criminal Courtroom,' *Journal of Legal History* 38, no. 2 (2017): 155.
22 Peter King, *Crime, Justice and Discretion in England, 1740–1820* (Oxford: Oxford University Press, 2000), 255.
23 Hone, *Three Trials.* All my page references to the trial are from William Tegg's 1876 reprint of Hone's publication, in which the pages are numbered consecutively. The transcripts in Hone's account and the newspapers are very close, mostly identical, but there are significant omissions in the newspaper's coverage. The precise source of Hone's fuller text has not yet been determined, but shorthand recorders were present in the court, as always. Joss Marsh suggests that Francis Place may have provided a shorthand writer for Hone's trials. See Joss Marsh, *Word Crimes: Blasphemy, Culture and Literature in Nineteenth-Century England* (Chicago: University of Chicago Press, 1998), 332 n.11.
24 E.P Thompson, *The Making of the English Working Class* (New York: Vintage, 1963), 721.
25 E.P. Thompson, *Customs in Common* (Harmondsworth: Penguin, 1993), 45.

26 Smith, *Politics of Language,* 177.
27 This is Hone's individual estimate of the number of people partly in the Hall and partly crowding the Avenues. See Routledge, *History of Popular Progress,* 434; David Kent and D.R. Ewen 'Introduction' in David Kent and D.R. Ewen, eds. *Regency Radical: Selected Writings of William Hone* (Detroit, MI: Wayne State University Press, 2004), 10. Hone's account of the second trial commenced in a similar narrative fashion: '... at a very early hour all the avenues of the Court were literally blocked up by a multitude of spectators, anxious to become auditors of the proceedings; and when the doors of the Court were opened not one-twentieth part of the multitude could find standing accommodation.' Hone, *Three Trials,* 71.
28 Kent and Ewen, 'Introduction,' 17.
29 Smith, *Politics of Language,* 196, also 197–98.
30 Smith, *Politics of Language,* 178, 189; on Hone's appeal to the Protestant martyr persona in light of Lilburne's precedent, see further Wood, *Radical Satire*, 121–33.
31 Hone, *Three Trials,* 23.
32 Hone, *Three Trials,* 55. Hone repeats in the second trial that 'libel is an undefined crime,' 95, and in the third trial states: 'as the law of libel stood at present, it was not possible to be understood ... (I)n fact, there was no such thing as law of libel; or, if there was, the law was written upon a cloud, which suddenly passed away, and was lost in vapour.' 154.
33 Hone, *Three Trials,* 55.
34 Hone, *Three Trials,* 86–87.
35 Rupert Cross and James W Harris, *Precedent in English Law,* 4th ed. (Oxford: Clarendon, 1991), 3.
36 See further Smith, *Politics of Language,* 154–201; Wood, *Radical Satire,* 96–154; Marsh, *Word Crimes,* 24–41.
37 Tuite, 'Not Guilty,' 41.
38 Hone, *Three Trials,* 45.
39 Hone, *Three Trials,* 59.
40 *The Times*, 17 December 1817, 3.
41 Hone, *Three Trials,* 46.
42 Hone, *Three Trials,* 46.
43 *The Times*, 17 December 1817, 3.
44 Hone, *Three Trials,* 59.
45 Hone, *Three Trials,* 64.
46 Hone, *Three Trials,* 64. This part was omitted from *The Times* coverage.
47 Ben Wilson, *The Laughter of Triumph: William Hone and the Fight for the Free Press* (London: Faber and Faber, 2005) 237.
48 Hone, *Three Trials,* 70. As Kevin Gilmartin notes, 'the utopian dimensions of this communal denouement remind us that printed versions of these trials were narrative constructions, typically brought out by radical publishers to achieve a political effect.' Gilmartin, *Print Politics,* 124.
49 Kyle Grimes has traced out the circulation through this correspondence and concluded that within one month of the appearance of Hone's publications in London, Hone himself had ceased publication, but there were more than ten provincial printers – from Newcastle to Penzance, Sheffield to Boston – issuing their own editions of the works. Kyle Grimes, 'Spreading the Radical Word: The Circulation of William Hone's 1817 Liturgical Parodies,' in *Radicalism and Revolution in Britain 1775–1848: Essays in Honour of Malcolm I. Thomis,* ed. Michael T. Davis (London: Macmillan, 2000), 143.

50 The text that was the subject of the first trial, *The Late John Wilkes's Catechism,* took the catechism, the *Political Litany,* subject of the second trial, took the litany, a series of petitions, invocations or supplications recited by the clergy and responded to by the congregation, and the *Sincecurist's Creed,* subject of the third and final trial, took the Athanasian Creed, which affirms a belief in the Holy Trinity.
51 Cited in Grimes, 'Spreading the Radical Word,' 150.
52 Wood, *Radical Satire,* 114.
53 Grimes, 'Spreading the Radical Word,' 146. In this era, newspapers, pamphlets and magazines were available for multiple use in coffee houses and taverns and might be read aloud to groups who included illiterate or semi-literate people.
54 Iain McCalman, *Radical Underworld Prophets, Revolutionaries and Pornographers in London, 1795–1840* (Oxford: Clarendon Press, 1988), 117, 122, cited in Kent and Ewen, 'Introduction,' 12.
55 Grimes, 'Spreading the Radical Word,' 155.
56 Grimes, 'Spreading the Radical Word,' 155.
57 Grimes, 'Spreading the Radical Word,' 154–55.
58 Mikhail Bakhtin, *Rabelais and his World* (Bloomington: Indiana University Press, 1984).
59 Grimes, 'Spreading the Radical Word,' 155.
60 Hone, *Three Trials*, 4.
61 For a close reading of this scene see Joseph M. Butwin, 'Seditious Laughter,' *Radical History* Review 18 (1978): 18–19.
62 Hone, *Three Trials*, 5.
63 Hone, *Three Trials*, 75.
64 Stanhope, Philip Dormer, Earl of Chesterfield, *Lord Chesterfield's Letters to his Son and Others*, ed. R.K. Root (London: J.M. Dent, 1969), 49.
65 Vic Gatrell, *City of Laughter: Sex and Satire in Eighteenth-Century London* (New York: Walker & Co., 2007).
66 Gatrell, *City of Laughter,* 9.
67 Wilson, *Laughter of Triumph,* 238.
68 Cited in Kent and Ewen, 'Introduction,' 19.
69 Hone, *Three Trials,* 74.
70 Hone, *Three Trials,* 74.
71 Law Report, *The Times*, 19 December 1817, 2.
72 Hone, *Three Trials,* 74.
73 Hone, *Three Trials,* 83.
74 Hone, *Three Trials,* 83.
75 Hone, *Three Trials,* 84. See also Hone, *Three Trials*, 99: 'The interruption could be occasioned by no friend of his: whoever disturbed the Court was his bitter enemy; his friends could not so conduct themselves; and the noise could only proceed from some designing emissaries, who were anxious that he should be taken from the Court to a prison.'
76 Gilmartin, *Print Politics,* 126.
77 Hone, *Three Trials,* 60.
78 Hone, *Three Trials,* 104.
79 Hone, *Three Trials,* 26.
80 Hone, *Three Trials,* 89.
81 Hone, *Three Trials,* 90.
82 *The Times*, Saturday 20 December 1817, 2.
83 Hone, *Three Trials,* 90.
84 Hone, *Three Trials,* 99.

85 Hone, *Three Trials*, 99.
86 Christina Parolin, *Radical Spaces: Venues of Popular Politics in London 1790–1845* (Canberra: ANU Press, 2010), 225.
87 Hone, *Three Trials*, 134.
88 Wilson, *Laughter of Triumph*, 252.
89 Hone, *Three Trials*, 138.
90 Epstein, *Radical Expression*, 55. See Hone, *Three Trials*, 72 and 138 (noting that there were only seven jurymen in attendance for the third trial.)
91 Hone, *Three Trials*, 139.
92 Hone, *Three Trials*, 139.
93 Hone, *Three Trials*, 146–47.
94 Hone, *Three Trials*, 148.
95 Hone, *Three Trials*, 148.
96 Hone, *Three Trials*, 148.
97 Hone, *Three Trials*, 149.
98 Hone, *Three Trials*, 149–50.
99 Hone, *Three Trials*, 157.
100 Hone, *Three Trials*, 158.
101 Wood, *Radical Satire*128.
102 Hone, *Three Trials*, 14–17, 92–97, 149–53. Hone's account in the first trial includes an account of appearing before Lord Ellenborough after his arrest, and feeling giddy and applying for leave to sit, to which Ellenborough replied 'No!' 'and it was pronounced,' recounted Hone,

> with an intonation that might have been heard at the further end of the hall. This refusal, instead of making him sink on the floor, as he had before expected to do, had the effect that a glass of water on being thrown into his face would have had, and he felt perfectly relieved. At the same time, however, he could not help feeling contempt for the inhumanity of the judge.

See Hone, *Three Trials*, 15.
103 Hone, *Three Trials*, 154.
104 Hone, *Three Trials*, 163.
105 Hone, *Three Trials*, 175; *The Times*, Monday 22 December 1817, 2.
106 Hone, *Three Trials*, 134.
107 Linda Mulcahy, *Legal Architecture: Justice, Due Process and the Place of Law* (New York: Routledge, 2011), 84.

Bibliography

Primary sources

Bentham, Jeremy. 'Draught of a Code for the Organisation of the Judicial Establishment in France' (1790). In *The Works of Jeremy Bentham*. 11 vols, edited by John Bowring. Edinburgh: William Tait, 1838–43.

Habeas Corpus Suspension Act 1817 (57 Geo III, c 3).

Hone, William. *The Three Trials of William Hone, for Publishing Three Parodies; viz. The Late John Wilkes's Catechism, The Political Litany, and the Sinecurist's Creed; on Three Ex-Officio Informations, at Guildhall, London, on Three Successive Days*. London: William Hone, 1818. Reprinted 1876.

Times.

Routledge, James. *The History of Popular Progress Chiefly in Relation to the Freedom of and Trial By Jury 1660–1820*. London: Macmillan, 1876.

Secondary Sources

Bakhtin, Mikhail. *Rabelais and His World*. Bloomington: Indiana University Press, 1984.

Bourdieu, Pierre. 'The Force of Law: Toward a Sociology of the Juridical Field.' *The Hastings Law Journal* 38, no. 5 (1987): 805–53.

Butwin, Joseph M. 'Seditious Laughter.' *Radical History Review* 18 (1978): 17–34.

Cross, Rupert and James W. Harris. *Precedent in English Law*. 4th ed. Oxford: Clarendon Press, 1991.

Davis, Michael T. 'Prosecution and Radical Discourse During the 1790s: The Case of the Scottish Sedition Trials.' *International Journal of the Sociology of Law* 33, no. 3 (2005): 148–58.

Epstein, James. *Radical Expression, Political Language, Ritual and Symbol in England, 1790–1850*. Oxford: Oxford University Press 1994.

Gatrell, Vic. *City of Laughter: Sex and Satire in Eighteenth-Century London*. New York: Walker & Co., 2007.

Gilmartin, Kevin. *Print Politics: The Press and Radical Opposition in Nineteenth-Century England*. Cambridge: Cambridge University Press, 1996.

Goodrich, Peter. 'The Continuance of the Antirrhetic.' *Cardozo Studies in Law and Literature* 4, no. 2 (1992): 207–22.

Goodrich, Peter. *Languages of Law: From Logics of Memory to Nomadic Masks*. London: Weidenfeld and Nicolson, 1990.

Goodrich, Peter. 'Specula Laws: Image, Aesthetic and Common Law.' *Law and Critique* 2, no. 2 (1992): 233–54.

Graham, Clare. *Law: The Architectural and Social History of the English Law Court to 1914*. London: Routledge, 2003.

Grimes, Kyle. 'Spreading the Radical Word: The Circulation of William Hone's 1817 Liturgical Parodies.' In *Radicalism and Revolution in Britain 1775–1848: Essays in Honour of Malcolm I. Thomis*, edited by Michael T. Davis, 143–56. London: Macmillan, 2000.

Harling, Philip. 'The Law of Libel and the Limits of Oppression, 1790 to 1832.' *Historical Journal* 44, no. 1 (2001): 107–34.

Kent, David and D.R. Ewen. 'Introduction.' In *Regency Radical: Selected Writings of William Hone*, edited by David Kent and D.R. Ewen, 9–27. Detroit, MI: Wayne State University Press, 2004.

King, Peter. *Crime, Justice and Discretion in England, 1740–1820*. Oxford: Oxford University Press, 2000.

Lemmings, David. 'Introduction: Criminal Courts, Lawyers and the Public Sphere.' In. *Crime, Courtrooms and the Public Sphere in Britain, 1700–1850*, edited by David Lemmings, 1–21. Farnham: Ashgate, 2012.

Marsh, Joss. *Word Crimes: Blasphemy, Culture and Literature in Nineteenth-Century England*. Chicago: University of Chicago Press, 1998.

McCalman, Iain. *Radical Underworld Prophets, Revolutionaries and Pornographers in London, 1795–1840*. Oxford: Clarendon Press, 1988.

Milka, Amy and David Lemmings. 'Narratives of Feeling and Majesty: Mediated Emotions in the Eighteenth Century Criminal Courtroom.' *Journal of Legal History* 38, no. 2 (2017): 155–78.

Mulcahy, Linda. *Legal Architecture: Justice, Due Process and the Place of Law*. New York: Routledge, 2011.

Parker, James. *Acoustic Jurisprudence: Listening to the Trial of Simon Bikindi*. Oxford: Oxford University Press, 2015.

Parolin, Christina. *Radical Spaces: Venues of Popular Politics in London 1790–1845*. Canberra: ANU Press, 2010.

Smith, Olivia. *The Politics of Language, 1791–1819*. Oxford: Clarendon Press, 1984.

Stanhope, Philip Dormer, Earl of Chesterfield, *Lord Chesterfield's Letters to His Son and Others*. Edited by R.K. Root. London: J.M. Dent, 1969.

Thompson, E.P. *Customs in Common*. Harmondsworth: Penguin, 1993.

Thompson, E.P. *The Making of the English Working Class*. New York: Vintage, 1963.

Tuite, Clara. 'Not Guilty: Negative Capability and the Trials of William Hone.' In *Censorship and the Limits of the Literary: A Global View*, edited by Nicola Moore, 33–48. London: Bloomsbury Publishing, 2012.

Wilson, Ben. *The Laughter of Triumph: William Hone and the Fight for the Free Press*. London: Faber and Faber, 2005.

Wood, Marcus. *Radical Satire and Print Culture 1790–1822*. Oxford: Clarendon Press, 1994.

9 Rioting and writing voice in Dickens's *Barnaby Rudge*

Helen Groth

In 1836, Charles Dickens chose the Gordon Riots of 1780 as the subject for his first foray into serious historical fiction, *Barnaby Rudge*. Remembered as a shameful outpouring of Protestant intolerance against Catholics that temporarily transformed London's streets into a violent conflagration, the Gordon riots captivated Dickens. Writing excitedly to his friend John Forster, Dickens describes being consumed by the riot's energy and momentum:

> I have let all the prisoners out of Newgate, burnt down Lord Mansfield's and played the very devil. Another number will finish the fires and help us on towards the end. I feel quite smoky when I am at work. I want elbow-room terribly.[1]

Running over multiple numbers in its initial serialised publication, the riot scenes in *Barnaby Rudge* take on a life of their own in Dickens's account, providing the necessary momentum to bring the novel to its climactic denouement. This letter reveals how formally central the riot was to *Barnaby Rudge*, but also suggests that Dickens struggled to structurally harness the energy it unleashed. His deeply felt desire for 'elbow room' evokes an image of the writer having to push through the noisy crowd of characters he has brought to life on the page. Liberated from Newgate the rioters run amok, diffusing the central focus of the novel, which is to bring together two very different plots – a detective plot in which the hapless Barnaby Rudge and his mother elude the murderous Barnaby Rudge senior only to find themselves swept up in the other parallel plot of the novel that dramatises the underground machinations of the Gordon Rioters, who burst out onto the streets of the novel and sweep up both mother and son, just as they do the author himself.

This chapter will argue that Dickens's struggle to contain the voices of the Gordon rioters in *Barnaby Rudge* tests the limit of his phonographic ambitions for the novel more generally. Transcribing the voices of the rioters and the noise of the riot is in this sense not incidental but pivotal to understanding Dickens's struggle to balance the demands of literary and social representation. This was a struggle that Dickens shared with many of his contemporaries, who were also drawn to the sheer dynamism and symbolic power of the riot as a form of spectacular instantiation of unchecked dissent. Disraeli, Bronte, Gaskell, Eliot, Kingsley and James all, in varying ways, mobilised different modes of rioting to explore the limits of their own and the reader's sympathetic engagement with a selected cause, or an individual rioter.

The realist novel provided a unique space for these writers to express the profound social anxieties that the chaotic vocality of the riot sparked and heightened, whilst simultaneously asserting fiction's role as a powerful alternative form of knowledge and social representation.[2] Taking inspiration from sound studies, this chapter emphasises how central the construction of voice and sound is to Dickens's reanimation of what Jonathan Sterne calls the 'audible past.'[3] Continually reminding the reader to listen to individual voices, rather than passively registering the noise of voices raised in chaotic unison, Dickens's novel transforms the reader into an historical 'ear-witness.'[4] Capturing idiosyncratic speech patterns is central to Dickens's evocation of character; indeed, one could argue that his fiction habituates a metonymic substitution of voice for character. Voice in this sense indicates a complex multimodality that shifts from speech to cognition to affect, and from character to narrator and back again, whilst subtly asserting the novel's capacities as a literal archive of vocal and acoustic traces.

Phonographic writing, *Barnaby Rudge* and the Gordon Riots

Dickens, as Ivan Krielkamp observes, wrote phonographically well before Edison invented the phonograph.[5] Phonography was a synonym for the shorthand transcription of voice from 1837, when Isaac Pitman coined the word to describe his new shorthand system. To write phonographically, prior to the invention of the phonograph, meant to capture the voice on paper, to provide a verifiable document of what was spoken or how an event unfolded. Few would disagree that Dickens's approach to the novel mimicked these documentary impulses, and yet, the question of whether a novel can be read as a document that bears witness to an event raises issues of both genre and format. Novelistic description necessarily structures, selects and filters in ways that the phonographic transcription of a voice does not. Correspondingly, the physical medium of the book or serial number of a journal, such as Dickens's *Household Words*, is necessarily distinct from a shorthand document transcribed on a sheet or sheets of paper.[6] What unites Dickens's version of the nineteenth-century novel and the shorthand documentation of an event – a riot, for example – is a mutual investment in information and authorised eyewitness accounts that attempt to solidify a particular version of events.

In the case of a fictional version of a riot scene, a novel may remediate documents, such as newspaper accounts or eyewitness testimony, lifting and poaching the rhetoric of reportage and transcription as part of the accumulative detailing of an alternative real, thus producing a genre or mode of writing that persistently unsettles the boundary between forms of knowledge and inscription. As Bruno Latour reminds us, 'anything and everything' can and has been transformed into inscription or 'paper writing' regardless of the genre or epistemology.[7] However, innovations in 'graphism,' as Latour is quick to qualify, are only crucial 'insofar as they allow new two-way relations to be established with objects (from nature or from fiction) and only insofar as they allow inscriptions either to become more mobile or to stay immutable through all their displacements.'[8] In the case of *Barnaby Rudge*, in the early 1830s, Dickens chose a well-documented historical event, the Gordon Riots of 1780, as the source for his move into serious historical fiction. Much had been written on the Gordon Riots in the intervening years, an accumulation of inscription that had generated a relatively consistent set of seemingly immutable facts, rendered portable and available for fictional hybridisation by their transformation into 'paper knowledge,' to use Lisa Gitelman's term.[9]

Writing as an eyewitness, without the advantage of retrospect, Horace Walpole's account of the riots vividly conveys the instability of immediate impressions transcribed in the heat of the moment:

> It is all unaccountable, and I can send you no consistent narrative. Much appears to have been sudden fury, and in many places the act of few. In other lights it looks like plan and deep premeditation. Whether it will ever be unravelled, I know not; or whether, like the history of dark ages, falsehood will become history, and then distant periods conjecture that we have transmitted very blundered relations: but, when I know so little of what has passed before my own eyes. I shall not guess how posterity will form their opinions.[10]

Diffused and fragmented by shock, Walpole's version of events is personal and local, one of many eyewitness accounts that had not, as yet, assumed the immutability of a generally known 'truth' materialised by multiple print copies, circulated and accumulated across time and space. Dickens, by contrast, writing in 1836, recreates the key events of the Gordon Riots with the confidence of assumed and well-documented knowledge. While this claim to mastery is not unusual for Dickens, there are telling points of narrative strain in the climactic moments of the novel that reveal his struggle to capture and channel the chaotic energies of the riot into a symbolic moment of political and social resolution, no matter how illusory, in which all voices are heard but the wisest prevail.

One version of what Walpole might have seen unfolding can be recounted as follows. On 2 June 1780, around 50,000 people sporting blue cockades gathered on St George's Fields, just south of Westminster Bridge in response to calls from the Protestant Association led by Lord

George Gordon to march on Parliament in support of a petition to repeal the Catholic Relief Act of 1778, which had lifted some restrictions on the civil rights of Roman Catholics, including removing the condemnation of Catholics from the oath of allegiance to the British Crown, as well as restrictions on land ownership, publishing and preaching the faith. While initially peaceful, the crowd was too large to remain so. Divided into four groups behind distinct flags and banners, the crowd followed four different routes to the Houses of Parliament. One of the groups, the London Scottish, was notably led by a Highlander in full tartan armed with a broad sword and accompanied by two pipers. Newspaper reports of this phase of the protest stress the civil behaviour of the marchers and the rousing popular support they received from the gathered spectators. The *St James's Chronicle* reported the following day:

> Though the Appearance of the Associators in General, was such as bespoke them Mechanicks and working people, yet was their behaviour at first, and during all their march through the City, such as would have done any Cause, any Rank, the highest Honour.[11]

This celebratory mood changed, however, by the time the protestors started to crowd into the streets around the Parliament. Coinciding with the arrival of members for the afternoon parliamentary session, the crowd took umbrage at perceived slights as the members' carriages forged their way through the crowded streets. Some protestors broke into the lobby and Lord Gordon demanded the petition signed by 44,000 signatories be dumped on the floor of the chamber and then preceded to address the crowd from the gallery of the chamber.

Resisting these provocations, the House of Commons adjourned further debate of the petition until the following Tuesday. It was this decision that sparked the chaos and violence that Walpole described. Some sections of the crowd dispersed, but those that remained were angered by the deferral of Gordon's motion. Late that night the riots began with the burning down of the Sardinian Embassy in Lincoln's Inn fields, the first of many buildings associated with Catholicism to be plundered and destroyed. Over the next six days, chapels and private houses of perceived Catholic sympathisers and prominent Catholics were attacked, as well as sites of state power such as prisons and the Bank of England. When Parliament reconvened on Tuesday, it again refused to debate Gordon's petition, but by this point Gordon himself was no longer in control of the revolutionary impulses of the crowd he had inspired.

On the night of Tuesday, 6 June, London was ablaze with fires lit by the rioters as they rampaged through the streets, burning the houses of prominent members of Parliament and the judiciary before moving on to Newgate prison. What ensued at Newgate was reported in both print and images in vivid detail, layers of inscription that became a matter of 'immutable' record by the time Dickens reanimated these events as the dramatic centrepiece of *Barnaby Rudge*. The crowd attacked Newgate, liberated its prisoners and burned it to the ground, the first of a series of climactic moments that culminated with 'Black Wednesday' and the attacks on the Bank of England. By this point, the King effectively proclaimed martial law, ordering thousands of troops into central London to shoot at the rioters without the constraint of reading the Riot Act. On the evening of that day, George Gordon, after failing to quell the rioters, joined the fight against them. The ensuing fatalities may have reached 700, but the exact figure is unknown. Gordon was arrested and charged with treason but was ultimately acquitted and released.

Dickens erroneously claims in the preface to the serial version of the novel published in *Master Humphrey's Clock* between February and November 1841 that 'No account of the Gordon Riots, having been to my knowledge introduced into any Work of Fiction, and the subject presenting very extraordinary and remarkable features, I was led to project this Tale.'[12] Despite this assertion, fictional accounts of the riots had provided a means of exorcising the spectre of the French Revolution in George Walker's *The Vagabond* (1799), Maria Edgeworth's *Harrington* (1817) and Thomas Gaspey's *The Mystery* (1820). As Miriam L. Wallace has argued, these novels contributed to the solidification of the historical perception of the Gordon Riots as integral to the 'revolutionary worldview' that emerged in the 1790s and 1800s. 'The 1780 riots become a kind of fulcrum,' Wallace argues, 'pointing back on the one hand to traditions of carnivalesque popular revolt, and forwards on the other towards a nineteenth-century concept of mass protest and revolutionary action by the proletariat.'[13]

Whether Dickens knew of these fictional precursors or not, this investment in historical precedence is consistent with both the hybrid form of *Barnaby Rudge* and its role as a filtering medium that made some voices more historically audible and sympathetically compelling than others. A more pointed and selective amnesia characterises Dickens's unacknowledged debt to Walter Scott, from whom he draws his historical conception of eighteenth-century England, prior to the riots, as a time of stasis and arrested development, typified by the habitués of John Willet's Maypole Inn, who smoke and drink for hours on end in a smug silence in which 'none of the company … pronounced one word.'[14]

The novel's subtitle 'A Tale of the Riots of "80",' if we accept the Oxford Dictionary definition of the genre of the tale, can be read as indicating either a fictitious or true narrative or story. The first preface favours fact over fiction, making a case for the veracity of Dickens's version of the riots based on his reliance on written documents, newspapers and eyewitness accounts and making no reference to the novel's equally central murder plot:

> It may be observed that, in the description of the principal outrages, reference has been made to the best authorities of that time, such as they are; and that the account given in this Tale, of all the main features of the Riots, is substantially correct.[15]

One of these 'authorities' was Thomas Holcroft (1745–1809), a self-taught cobbler's son. Published under the pseudonym 'William Vincent, Esq., of Gray's Inn,' *A Plain and Succinct Narrative of the Late Riot and Disturbances in the Cities of London and Westminster, and the Borough of Southwark* (1780) is anything but plain and succinct.[16] Holcroft's account is a syncretic transcription of diverse voices, perspectives and sources, the latter including a summarised version of the Catholic Relief Bill, George Gordon's advertisements for the Protestant Association petition, the King's proclamation calling in the militia to quell the riot, various contemporary handbills, newspapers, transcribed rumours and eyewitness accounts. These sources are combined into a 'narrative' that blurs the generic lines between historical document and fictional tale, or storytelling. As this description suggests, Holcroft's account was ultimately driven by an impulse to both enact and contain the chaotic vocality of the riots; to create a 'new inscription,' to invoke Latour, which would have 'the properties of being mobile but also immutable, presentable, readable and combinable with another.'[17]

Like Dickens, Holcroft was a novelist, a journalist and a dramatist from a humble background, although his politics were more radically inclined than Dickens's quietist liberal reformism. Accordingly, Holcroft's account oscillates between anxiety that the rioters are being manipulated by a sinister charismatic influence and a more sympathetic recognition of the natural energy driving the mob's desire for political recognition and reciprocity. At one point, he speaks of a 'wicked head' directing the riot:

> The mischief which had been perpetrated, and that which was evidently intended, were so black an aspect, and involved so entirely the destruction of empire, that the most dispassionate, and those who were least liable to be perverted by idle conjecture, could not help sometimes thinking they saw a wicked head directing the violent and rude hands of the mob.[18]

Counterbalancing this conspiratorial version of events, Holcroft reflects on the natural energy of the crowd in his description of the destruction of Newgate:

> It is almost incredible to think that it were possible to destroy a building of such amazing strength and extent, with so much swiftness as

> they accomplished this ... The activity of the Mob was in this instance, as well as in every other, amazing. They dragged out the prisoners, many of them, by the hair of the head, by the legs or arms, or whatever part they could lay hold of: They broke open the doors of the entrances, as easily as if they had all their lives been acquainted with the intricacies of the place, to let the confined escape ... so well planned were all the manoeuvres of these desperate ruffians, that they placed centinels at the avenues, to prevent any of the prisoners from being conveyed to other jails.[19]

As Miriam L. Wallace has argued, this oscillation between anxiety and grudging admiration in Holcroft's narrative draws on two dominant eighteenth-century theories of the mob that Ronald Paulson identifies in accounts of the Gordon Riots – the Tory and the Whig mob.[20] In the Tory version of the mob, there is always a charismatic manipulative leader coercing and driving the action with sinister intent. By contrast, the Whig mob is driven by natural force and human impulse to embrace freedom of speech and action, to quote Paulson:

> The Enlightenment was the age of 'Natural Law,' and the riot could be seen as an attack on a wrong that, on the principle of 'natural rights,' should and will be righted. The riots were about restraint and the breaking of restraints, and so invoked the metaphor of ungovernable natural forces of fire and water. Thus the political right viewed the riot as a perversion of nature and the left as a natural phenomenon.[21]

Traces of Holcroft's ambivalence towards the crowd can be detected in Dickens's sympathetic treatment of 'the idiot' Barnaby Rudge, who is both caught up in the oceanic flow of the mob and manipulated by the uncivil brutishness of Hugh, the illegitimate son of the Protestant Sir John Chester. The malign plotting of Gordon's Machiavellian acolyte Gashford, as well as the criminal actions of his own father which are the catalyst for the novel's overarching murder plot, likewise recall Holcroft. These affinities aside, recognising the generic difference between the form of Holcroft's account and the multivocal or dialogic form of the realist novel is also fundamental to understanding how the complex vocality of the rioters is registered in *Barnaby Rudge*. Where Holcroft mediates 'the hisses, groans, and hootings' and the collective cries of 'No Popery, no Popery!'[22] as one of many forms of data that comprise his historical narrative, Dickens's digressive and detailed re-enactment of the Gordon Riots is filtered through the two definitive and contradictory energies of the realist novel – its democratic impulse to represent the many and its aesthetic reproduction of social inequities. To quote Alex Woloch,

> The realist novel is infused with the sense that any character is a potential hero, but simultaneously enchanted with the freestanding

> individual, defined through his or her interior consciousness. In the paradigmatic character-structure of the realist novel, any character *can* be a protagonist, but only one character is: just as increasing political equality, and a maturing logic of human rights develop amid acute economic and social stratification.[23]

Capturing the eccentric particularity of individual voices, their distinct rhythms and cadences is one of Dickens's key techniques for creating space for individual characters, both central and minor. In *Barnaby Rudge*, the simple passionate speech of the novel's titular character is rarely heard amidst the hubbub created by the multiple voices of minor characters competing for space in the crowded interiors and street scenes that dominate the novel. But on the odd occasions when Barnaby does speak, his child-like phrasing creates a space for a more sympathetic reading of the potential idealism and enthusiastic naiveté of individual rioters unwittingly drawn into the political machinations of an ambitious few.

Nowhere is this disparity between voices more clearly marked than in the description of Barnaby 'stationed' at the head of the rioters, where he walks between his supposed friend Hugh, 'that dangerous companion,' and Dennis the hangman:

> Forgetful of all other things in the ecstasy of the moment, his face flushed and his eyes sparkling with delight, heedless of the weight of the great banner he carried ... the only light-hearted, undesigning creature, in the whole assembly.[24]

Barnaby's inarticulate joy and failure to hear the real meaning of his fellow rioters' words recalls Carlyle's contemporaneous characterisation of the Chartists as 'wild inarticulate souls.'[25] Yet, unlike Carlyle, who only allows his crowd of 'dumb creatures' to speak in one voice as they pray for heroic leadership – 'Guide me, govern me! I am mad and miserable, and cannot guide myself' – Dickens gives Barnaby direct speech of his own.[26] Throughout the riot, Dickens uses Barnaby's comments to register a counter-discourse of unalloyed loyalty and bravery. Barnaby assures his companions that 'Nobody will wrest this flag away' in the midst of the 'execrations, hoots and howlings' of the 'mad monster' that the mob had become.[27] Contrasting with the increasing confusion of Gordon whose speech, the narrative voice informs the reader, had become 'as childish, irresolute and uncertain' in the face of the 'noise and exasperation' of the rioters, Dickens consistently uses Barnaby's voice in dialogue and working at cross purposes with the violent intent of his companions Hugh and Dennis.[28] In response to Hugh's insincere inquiry after his wellbeing at a critical moment when the riot is spinning out of control – 'How are you, boy?' – Barnaby's response registers the gulf between his experience of the riot and the brutal religious intolerance that drives Gashford and his acolytes, Hugh and Dennis:

> 'Hearty!' cried Barnaby, waving his hat. 'Ha, ha, ha! And merry too Hugh! And ready to do anything for the good cause, and the right, and to help the mild, pale-faced gentleman – the Lord they use so ill – eh, Hugh?'[29]

This dissonance between Barnaby's simple exclamations and the complex sophistry and 'violent words' of Gashford, the key advisor of Gordon, 'the mild, pale-faced gentleman,' introduces the idea of mishearing at this critical moment in the plot while implicitly raising the possibility that the crowd is too confused and multivocal to be understood as speaking in one voice.[30]

Even Gordon is given some room to move in the space created by the parallels between Barnaby's manipulation and his own transition from central agent to political pawn, both engineered by the conspiratorial Gashford: a sympathetic, if muted, recasting of the role of the riot's chief instigator and ultimate traitor that some of Dickens's contemporaries, including his biographer and friend John Forster, found perturbing.[31] It is worth noting, however, that this gesturing towards the possibility of a more democratic representation of the complex social conditions underlying the personal history of individual rioters strikes a minor note in a novel ultimately driven by a high Victorian liberal argument for religious tolerance, as well as an anti-chartist inspired assertion of the necessary triumph of rational order in the face of radical insurgency.[32] The latter is literalised in the novel's systematic calculation of the evidence of the crimes of the rioters, the number of dead, wounded, imprisoned and injured parties, after 'the crowd [is] utterly routed.'[33]

Sympathetic listening, distributed voices and social space

Barnaby dwells with his mother in the relative quiet of the countryside just outside London for most of the first half of the novel. There they live in genteel poverty not far from the declining estate, the Warrens and the convivial, if stifling, charms of the Maypole Inn. Far from idyllic, their present is haunted by the past actions of Barnaby Rudge senior, who is implicated and is indeed guilty of the murder of his master, the onetime owner of the Warrens, Reuben Haredale. Forced to find refuge in the protective anonymity of London's crowded streets when Barnaby Snr unexpectedly returns, son and mother, the latter having carried the secret of her husband's guilt all along, are unwittingly drawn into the noise and chaos of the Gordon Riots; indeed, Barnaby meets Gordon himself on the day he arrives in London in a typical instance of Dickensian coincidence. It is at this midpoint in the novel that the two plots that enact parallel challenges to traditional forms of social order converge – the murder plot in which a servant kills a master and the political plot in which the rioters challenge the dominant political order. The relative narrative dominance of the local tale-telling and information sharing carried on by the benign village patriarchs that crowd the bar at the Maypole also gives way at this juncture to the orchestrated intelligence and secret activities typified by the far more sinister denizens of The Boot, the tavern where the rioters meet to plot the events that unfold in the climactic moments of the novel.

What is striking about the scenes that describe the early phases of the riot is the movement between dialogue, reported speech and free indirect discourse in which the narrative voice both individualises and blends the secret meetings, as well as the cries, shouts and huzzahs of the rioters as they surge through the streets of London. Some voices are sanctioned, others are caricatured, such as Sim Tappertit, the leader of the 'Prentice Nights' who admits new members to the subterranean company he captains with a confusing meld of radical proselytising and delusional self-aggrandisement that is parodied by the narrative voice that reports the content of his speech:

Having said it, he listened meekly to the captain, who, in an address prepared for such occasions, told him how that under that same Constitution (which was kept in a strongbox somewhere, but where exactly he could not find out, or he would have endeavoured to procure a copy of it), the 'prentices had, in times gone by, had frequent holidays of right, broken people's heads by scores, defied their masters, nay, even achieved some glorious murders in the streets, which privileges had gradually been wrested from them, and in all which noble aspirations they were now restrained; how the degrading checks imposed upon them were unquestionably attributable to the innovating spirit of the times, and how they united therefore to resist all change, except such change as would restore those good old English customs, by which they would stand or fall.[34]

Blurring the lines between the voice of Sim and the narrative voice, this passage transgresses the boundary between character, plot and narrative process. Traces of Sim's address to the young novices and the argument for constitutional representation that it mediates are channelled and rendered both absurdly nostalgic and structurally subordinate by the accumulative parodic energy of the narrative voice. This creates a transient space for Sim to speak and then shuts it down just as unceremoniously in the rush to move on to the next minor character and scene as the novel builds up the complex network of forces and agents that feed into the climatic events of the Gordon Riots. In this instance, the constitutional issues addressed by the Gordon Riot petition are also confused with the associative noise generated by a more immediate contemporary anxiety about the potential constitutional crisis and revolutionary potential of the secret gatherings of the Chartists in the 1830s and 1840s, as Brantlinger and Marcus have observed in their influential readings of the novel.[35] This more present political noise also distracts from the underlying point of this particular parody, which is Dickens's interest in prising the rhetoric of intolerance and irrationality apart from the collective understanding of a progressive selectively representative English political system. Sim Tappertit in the context of the overall plot is the irrational chaotic antithesis of the novel's principal embodiment of responsible citizenship and rational empathy, Sim's employer, Gabriel Varden.

Gabriel Varden, as the narrative voice reminds the reader, is one of the few who listens closely to Barnaby and his kind, and, in the process, remains attuned to the sounds and voices of the street, unlike the 'men of gloom and austerity' whose reformist intolerance makes them deaf to the 'music,' 'songs and cheerful sounds' of the many.[36] Breaking frame to address this anonymous judgemental minority, the narrative voice urges:

> Listen to the million voices in the summer air, and find one dismal as your own. Remember, if ye can, the sense of hope and pleasure which every glad return of day awakens in the breast of all your kind who have not changed their nature; and learn some wisdom even from the witless, when their hearts are lifted up they know not why, by all the mirth and happiness it brings.[37]

A few chapters later, Gordon and his malign counsellor Gashford listen very differently to the voices of the crowd, a distinction that the narrative voice invites the reader to attend to closely by carefully distinguishing between the two men's definitive characteristics before allowing the reader to listen in on their private conversation about the popular reception of one of Gordon's many inspirational calls to action. Gordon, 'the great personage,' is described as of middling height, slender, sallow, with an aquiline nose and 'long hair of reddish brown,' dressed plainly in an unadorned 'full suit of black.'[38] The narrator focusses the reader's attention on 'his very bright large eye, which betrayed a restlessness of thought and purpose' at odds with his otherwise sober mien and 'quaint and sad apparel.'[39] In contrast to Gashford, who is described as 'sly and slinking,' the narrator insists that Gordon's face had 'nothing harsh or cruel in its expression,' only a prevailing melancholy and 'indefinable uneasiness' that elicited sympathetic attention from others.[40] This sympathetic framing is reinforced by the coercive role attributed to Gashford in the ensuing conversation, in which the latter manipulates Gordon from reflective unease to delusional hubris in the interests of pursuing his own agenda rather than that of the public good – an invidious outcome foreshadowed by the narrator's encapsulation of his predatory character: 'He wore the aspect of a man who was always lying in wait for something that *wouldn't* come to pass.'[41]

Gordon's voice steadily increases in volume and fervour as he moves, cued by Gashford's sophistic flattery, from doubting his capacity to move the crowd, to a feverish declaration that they

> are the only men who regard the mass of people out of doors, or are regarded by them, we will uphold them to the last; and will raise a cry against these un-English Papists which shall re-echo through the country, and roll with a noise like thunder.[42]

This transformation of Gordon's voice from a halting unease in which traces of reason and humanity are still audible to the violent assertion of irrational passion that breaks free of all restraint and becomes inexplicable violent noise also implicitly parallels the violent emotion that will consequently ignite the 'mass of people out of doors.'[43]

At this point, the narrative voice steps back to comment on the significance of this transformation, creating a space to consider the effects of Gashford's malign charismatic influence:

> It would be difficult to convey an adequate idea of the excited manner in which he gave these answers to the secretary's promptings; of the rapidity of his utterance, or the violence of his tone and gesture; in which, struggling through his Puritan's demeanour, was something wild and ungovernable which broke through all restraint.[44]

Once Gordon's voice becomes consumed by violence and emotion, it is increasingly aligned with noise and 'ungovernable' affect. The resonant dimensions of voice in this instance press against the formal limits of transcription. It is, as the narrator observes, impossible to provide a literal phonographic recording of the emotional charge and tone of Gordon's passionate expression. The reader in this instance must settle for a descriptive enactment that blurs sight and sound, reading and listening.

In the ensuing pages, the narrative voice describes Gordon as 'the mere creature of his disordered brain,'[45] who few would have followed without the stimulus of 'vague rumours' of the rise of Popish powers intent on degrading and enslaving England.[46] Caught up in an oceanic tide of contagious fear and 'mania,' Gordon's voice is channelled and amplified by the mass distribution of a radical literature designed to haunt the collective imaginary of 'the ignorant and credulous' with the absurd prospect of 'an inquisition in London' that would transform 'the pens of Smithfield market into stakes and cauldrons.'[47] Implicit again here is Dickens's more immediate and anxious registration of the revolutionary momentum driving Chartism's effective use and distribution of popular print forms to build support for 'the disturbances of 1839,' as Ian Haywood has noted.[48] Pamphlets are 'dropped in the public ways, thrust under the house-doors, tossed in at windows and pressed into the hands of those who trod the streets by night,' while inflammatory posters

> glared from every wall, and shone on every post and pillar, so that stocks and stones appeared infected with the common fear, urging all men to join together blindfold in resistance of they knew not what, they knew not why.[49]

Just like the present disturbances created by the dangerous words of a radical minority, the appeal to the acoustic and visual evokes sensory overload here and, ideally, the correspondent impulse to shut down and filter out the noise generated by the counter-public sphere of Gordon's revolutionary movement and its contemporary iteration in the uncivil form of Chartist dissent. Acoustic metaphor mobilised in this way operates as a foreclosing hermeneutics that filters out the unwanted animus of an anachronistic revolutionary past. Aptly one of the actual names of the pamphlets circulated by Gordon's followers was *The Thunderer.*[50] Dickens describes this pamphlet being 'read aloud, to an eager knot of listeners,' while others whispered in corners 'jealous of their speech being overheard.'[51] Both modes of listening are judged harshly by the narrative voice.

In a brief lull before the riot begins in earnest, the narrative provides another exemplary model of a more civilised and rational form of listening to the authentic voices and rhythms of working London. In the locksmith Gabriel Varden's workshop, the sounds of honest labour transform the jarring discord of London into a melodic refrain:

> Women scolded, children squalled, heavy carts went rumbling by, horrible cries proceeded from the lungs of hawkers; still it struck in again, no higher, no lower, no louder, no softer; not thrusting itself on people's notice a bit the more for having been outdone by louder sounds – tink, tink, tink, tink, tink.
>
> It was a perfect embodiment of the still small voice, free from all cold, hoarseness, huskiness, or unhealthiness of any kind; foot passengers slackened their pace, and were disposed to linger near it; neighbours who had got up splenetic that morning, felt good humour stealing on them as they heard it, and by degrees became quite sprightly; mothers danced their babies to its ringing; still the magical tink, tink, tink, came gaily from the workshop of the Golden Key.[52]

This passage invites the reader to turn away from unreflective noise, to move inward, to attend closely to 'the still small voice' and the minor rhythmic patterns created by the 'tink, tink, tink' of Varden's hammer. What is being required here of both Dickens's fictional listeners and the reader is an attention to the pure form and the transcendent value of individual labour, of both writer and blacksmith. Both are beautiful, idealised forms of work, divested of the brutal excesses of political speech and the banalities of everyday verbal expression.

Jean-Luc Nancy draws attention to the 'pious ring' that can characterise such seemingly disinterested listening, confusing the passivity of 'tuning in' to the minor sounds of Gabriel's hammer with a more active engagement with the obscure lives of London's working poor.[53] By making the 'tink, tink, tink' of the hammer momentarily audible, transiently present but in such a minor way, the subsequent fall into silence of this resonant sound is inextricably linked to its distorted significance at this transitional moment in the plot. Peculiarly static, this moment promises more revelatory significance than it delivers: a fitting prelude to the ensuing riot scenes and the oscillation between momentarily audible obscure voices and the chaotic vocality of the crowd that ultimately subsumes them. Barnaby's naïve utterances throughout these scenes register the inevitable failure of the rioters' demand for representation and reciprocity. Deaf to the confusion of cries from 'honest zealots' and the cynical obscenities of 'the very scum and refuse of London,'[54] Barnaby assures his Machiavellian companions 'with glistening eyes' that he will hold the banner tight, while scanning the crowd for his mother's face: 'Would it make her glad to see me at the head of this brave show?'[55] Drawing the reader's attention to the theatricality of the riot through the distorted lens of Barnaby's sentimentalism invites a passive reception of a rapid succession of violent scenes in which Barnaby's failure to hear or understand what is really happening sweeps him up in one of the most spectacular and brutal events of the riot, the burning of the Holborn brewery. Appearing at the front of a crowd in which participants are covered in soot, 'their garments torn to rags; their hair hanging wildly,' 'their hands and faces jagged and bleeding,' Barnaby and his companions resemble 'hideous madmen' amidst the 'dense throng' of singing, shouting and quarrelling rioters.[56] Intensifying the horror of this spectacle still further the narrative voice summons the familiar animus of the Tory mob:

> a vision of coarse faces … a dream of demon heads and savage eyes … a bewildering horror, in which so much was seen, and yet so little, which seemed so long and yet so short, in which there were so many phantoms, not to be forgotten all through life, and yet so many things that could not be observed in that distracting glimpse – it flitted onward, and was gone.[57]

Jarring the reader out of this lurid phantasmagoria is a single 'piercing scream' that draws attention momentarily to the individual cost of this human travesty in the form of 'a widowed woman' who 'had descried her son among the rioters.'[58] In a cruel plot twist, Barnaby ultimately gets his wish, his mother witnesses his illusory central role in 'this brave show' and the listening reader is rewarded with the privilege of comprehending the true meaning of her scream, unlike the perplexed onlookers caught up in the riot's 'wrath and ruin.'[59]

From this point, the narrative's evocation of the noise and chaos of the riot converges with the rhetoric of contagion and distorted affect building towards the final climactic sacking of Newgate. Little space is given to the individual agency of rioters in the descriptive rush to evoke the sheer scale and potentially revolutionary horror of the 'moral plague' let loose in the city by Gordon and his acolytes, as the narrative voice observes with grim relish:

> The noise, and hurry, and excitement, had for hundreds and hundreds an attraction they had not firmness to resist. The contagion spread, like a dread fever: an infectious madness, as yet not near its height, seized on new victims every hour, and society began to tremble at their ravings.[60]

Dickens's compulsion to capture and record, to write phonographically in other words, in sequences like this encapsulates the complex paradox which the chaotic vocality of the Gordon Riots presents to his version of the realist novel. By placing Barnaby at the centre of the riot, as the gullible innocent who ultimately eludes the hangman's noose, Dickens creates a space for alternative voices and multiple perspectives in a novel which he claimed had a simple didactic purpose in his 1841 Preface: 'It is unnecessary to say, that those shameful tumults, while they reflect indelible disgrace upon the time in which they occurred, and all who had act or part in them, teach a good lesson.'[61] Like the 'tink, tink, tink' of Gabriel Varden's hammer, understanding the meaning behind Barnaby's naïve speech requires a more active mode of listening that jars with Dickens's drive to silence the revolutionary possibility of the riot as a more generalisable contemporaneous form of popular protest by damning it by association with the religious intolerance and xenophobic violence that the Gordon Riots had come to symbolise by the early 1830s when he sat down to write *Barnaby Rudge*.

Notes

1 Dickens to John Forster, 18 September 1841, in *The Letters of Charles Dickens*, ed. Madeline House, Graham Storey et al., 12 vols. (Oxford: Clarendon Press 1965–2002) 2:385.
2 Richard Menke reads the transformation of the violent speech of radical politics into the muted speech of idealised male working-class authorship as a problematic consequence of the professionalisation of the literary field in 'Cultural Capital and the Scene of Rioting: Male Working-Class Authorship in *Alton Locke*,' *Victorian Literature and Culture* 28, no. 1 (2000): 87–108.
3 Jonathan Sterne, *The Audible Past: Cultural origins of Sound Reproduction* (Durham, NC: Duke University Press, 2003), i.
4 Raymond Murray Schafer, *The Soundscape: Our Sonic Environment and the Tuning of the World* (Rochester, Vermont: Destiny Books, 1977), 8.

5 Ivan Krielkamp, *Voice and the Victorian Storyteller* (Cambridge: Cambridge University Press, 2005), 30.
6 These distinctions between paper and book, document and novel are indebted to Lisa Gitelman's opening discussion of the specific genre of the document in *Paper Knowledge: Toward a Media History of Documents* (Durham, NC: Duke University Press, 2014), 5–7.
7 Bruno Latour, 'Drawing Things Together,' in *Representation and Scientific Practice*, eds. Michael Lynch and Steve Woolgar (1988; Cambridge, MA: MIT Press, 1990), 28.
8 Latour, 'Drawing Things Together,' 28.
9 Gitelman, *Paper Knowledge*.
10 Horace Walpole to Horace Mann, 14 June 1780, in *The Letters of Horace Walpole, Earl of Orford*, ed. Peter Cunningham, 9 vols. (London: Henry G. Bohn, 1861–66), 2:401.
11 *St James's Chronicle*, 1–3 June 1780, cited in Ian Haywood and John Seed, 'Introduction,' in *The Gordon Riot: Politics, Culture and Insurrection in Late Eighteenth-Century Britain*, eds. Ian Hayward and John Seed (Cambridge: Cambridge University Press, 2012), 3.
12 Charles Dickens, 'Preface to Barnaby Rudge,' in *Barnaby Rudge*, ed. Clive Hurst (1841; Oxford: Oxford University Press, 2008), 3.
13 Miriam L. Wallace, 'Thomas Holcroft and the Gordon Riots: Romantic Revisionings,' in *The Gordon Riots: Politics, Culture and Insurrection in Late Eighteenth-Century Britain*, eds. Ian Hayward and John Seed (Cambridge: Cambridge University Press, 2012), 163.
14 Dickens, *Barnaby Rudge*, 17.
15 Dickens, 'Preface to Barnaby Rudge,' 3.
16 William Vincent [Thomas Holcroft], *A Plain and Succinct Narrative of the Late Riot and Disturbances in the Cities of London and Westminster, and the Borough of Southwark* (London: Fielding and Walker, 1780).
17 Latour, 'Drawing Things Together,' 21.
18 Vincent [Holcroft], *Plain and Succinct Narrative,* 44.
19 Vincent [Holcroft], *Plain and Succinct Narrative*, 27–28.
20 Wallace, 'Thomas Holcroft and the Gordon Riots,' 166.
21 Ronald Paulson, 'Four American Riots,' *Hopkins Review* 2, no. 4 (2009): 570.
22 Vincent [Holcroft], *Plain and Succinct Narrative,* 17.
23 Alex Woloch, *The One vs the Many: Minor Character and the Space of the Protagonist in the Novel* (Princeton, NJ: Princeton University Press, 2003), 19.
24 Dickens, *Barnaby Rudge*, 390.
25 Thomas Carlyle, *Chartism* (1839), in *Selected Writings*, ed. Alan Shelston (London: Penguin, 1971), 155. John Plotz provides an excellent reading of voice in Carlyle's *Chartism* in *The Crowd: British Literature and Public Politics* (Berkeley: University of California Press, 2000), 127–53.
26 Dickens, *Barnaby Rudge*, 184.
27 Dickens, *Barnaby Rudge*, 392–93.
28 Dickens, *Barnaby Rudge*, 395.
29 Dickens, *Barnaby Rudge*, 423.
30 Dickens, *Barnaby Rudge*, 424.
31 John Forster, *Life of Dickens*, ed. J.W.T. Ley (London: Cecil Palmer, 1928), 170.
32 There has been considerable work on the novel as an anxious response to contemporary concerns about the potentially revolutionary rise of Chartism, including the following seminal discussions: Patrick Brantlinger, *Spirit of Reform: British Literature and Politics, 1832–1867* (Cambridge, MA: Harvard University Press, 1977), 91; Steven Marcus, *Dickens from Pickwick to Dombey* (New York, NY: Simon & Schuster, 1965), 172–74.

33 Dickens, *Barnaby Rudge,* 605. Maurice S. Lee draws attention to Dickens's evidence gathering in *Barnaby Rudge* in 'Evidence, Coincidence, and Superabundant Information,' *Victorian Studies* 54, no. 1 (2011): 87–94.
34 Dickens, *Barnaby Rudge,* 76.
35 Patrick Brantlinger, 'The Case Against Trade Unions in Early Victorian Fiction,' *Victorian Studies* 13, no. 1 (1969): 37–52; Marcus, *Dickens from Pickwick to Dombey,* 169–212.
36 Dickens, *Barnaby Rudge,* 201–202.
37 Dickens, *Barnaby Rudge*, 202.
38 Dickens, *Barnaby Rudge*, 283.
39 Dickens, *Barnaby Rudge*, 283.
40 Dickens, *Barnaby Rudge*, 283.
41 Dickens, *Barnaby Rudge*, 283.
42 Dickens, *Barnaby Rudge*, 287.
43 Dickens, *Barnaby Rudge*, 287.
44 Dickens, *Barnaby Rudge*, 287.
45 Dickens, *Barnaby Rudge*, 295.
46 Dickens, *Barnaby Rudge*, 294.
47 Dickens, *Barnaby Rudge*, 294.
48 Ian Haywood, *The Revolution in Popular Literature: Print, Politics and the People, 1790–1860* (Cambridge: Cambridge University Press, 2004), 231.
49 Dickens, *Barnaby Rudge,* 294.
50 Dickens, *Barnaby Rudge*, 315.
51 Dickens, *Barnaby Rudge*, 315.
52 Dickens, *Barnaby Rudge*, 326.
53 Jean-Luc Nancy, *Listening*, trans. Charlotte Mandell (Fordham: Fordham University Press, 2007), 4.
54 Dickens, *Barnaby Rudge,* 393.
55 Dickens, *Barnaby Rudge*, 391.
56 Dickens, *Barnaby Rudge*, 404.
57 Dickens, *Barnaby Rudge*, 404.
58 Dickens, *Barnaby Rudge*, 404.
59 Dickens, *Barnaby Rudge*, 404.
60 Dickens, *Barnaby Rudge*, 421.
61 Dickens, 'Preface to Barnaby Rudge,' 3.

Bibliography

Primary sources

Carlyle, Thomas. 'Chartism' (1839). In *Selected Writings*, edited by Alan Shelston. London: Penguin, 1971.

Dickens, Charles. *Barnaby Rudge* (1841). Edited by Clive Hurst. Oxford: Oxford University Press, 2008.

Dickens, Charles. *The Letters of Charles Dickens.* Edited by Madeline House, Graham Storey et al. 12 vols. Oxford: Clarendon Press, 1965–2002.

Vincent, Williams. [Holcroft, Thomas]. *A Plain and Succinct Narrative of the Late Riot and Disturbances in the Cities of London and Westminster, and the Borough of Southwark*. London: Fielding and Walker, 1780.

Walpole, Horace. *The Letters of Horace Walpole, Earl of Orford.* Edited by Peter Cunningham. 9 vols. London: Henry G. Bohn, 1861–66.

Secondary sources

Brantlinger, Patrick. 'The Case against Trade Unions in Early Victorian Fiction.' *Victorian Studies* 13, no. 1 (1969): 37–52.

Brantlinger, Patrick. *Spirit of Reform: British Literature and Politics, 1832–1867.* Cambridge, MA: Harvard University Press, 1977.

Forster, John. *Life of Dickens.* Edited by J.W.T. Ley. London: Cecil Palmer, 1928.

Gitelman, Lisa. *Paper Knowledge: Toward a Media History of Documents.* Durham, NC: Duke University Press, 2014.

Haywood, Ian. *The Revolution in Popular Literature: Print, Politics and the People, 1790–1860.* Cambridge: Cambridge University Press, 2004.

Haywood, Ian and John Seed. 'Introduction.' In *The Gordon Riots: Politics, Culture and Insurrection in Late Eighteenth-Century Britain*, edited by Ian Hayward and John Seed, 1–18. Cambridge: Cambridge University Press, 2012.

Krielkamp, Ivan. *Voice and the Victorian Storyteller.* Cambridge: Cambridge University Press, 2005.

Latour, Bruno. 'Drawing Things Together.' In *Representation and Scientific Practice*, edited by Michael Lynch and Steve Woolgar, 19–68. Cambridge, MA: MIT Press, 1990.

Lee, Maurice S. 'Evidence, Coincidence, and Superabundant Information.' *Victorian Studies* 54, no. 1 (2011): 87–94.

Marcus, Stephen. *Dickens from Pickwick to Dombey.* New York: Simon and Schuster, 1965.

Menke, Richard. 'Cultural Capital and the Scene of Rioting: Male Working-Class Authorship in *Alton Locke*.' *Victorian Literature and Culture* 28, no. 1 (2000): 87–108.

Nancy, Jean-Luc. *Listening.* Translated by Charlotte Mandell. Fordham: Fordham University Press, 2007.

Paulson, Ronald. 'Four American Riots.' *Hopkins Review* 2, no. 4 (2009): 560–84.

Plotz, John. *The Crowd: British Literature and Public Politics.* Berkeley: University of California Press, 2000.

Schafer, R. Murray. *The Soundscape: Our Sonic Environment and the Tuning of the World.* Rochester, VT: Destiny Books, 1977.

Sterne, Jonathan. *The Audible Past: Cultural Origins of Sound Reproduction.* Durham, NC: Duke University Press, 2003.

Wallace, Miriam L. 'Thomas Holcroft and the Gordon Riots: Romantic Revisionings.' In *The Gordon Riots: Politics, Culture and Insurrection in Late Eighteenth-Century Britain*, edited by Ian Hayward and John Seed, 162–81. Cambridge: Cambridge University Press, 2012.

Woloch, Alex. *The One vs the Many: Minor Character and the Space of the Protagonist in the Novel.* Princeton, NJ: Princeton University Press, 2003.

Part IV

Civil and uncivil sounds of empire

10 The civil noise of empire

Bruce Buchan

Introduction

On 4 June 1788, the infant colony in Sydney Cove was treated to a rare sonic celebration. The 'First Fleet' that brought the first convict 'colonists' to establish a permanent British colony in New Holland (Australia) had been at Sydney for just over five months. David Blackburn, the Master aboard HMS *Supply*, one of the two ships then at anchor, wrote that: 'No Cannon had ever been fired since our arrival on the Coast ... but on that day being "His Majesty's Birth Day," the two ships 'fir'd a salute of 21 Cannon Each – at Sun Rise, Mid Day and at Sun Set.'[1] Daniel Southwell, a deck officer of HMS *Sirius*, who seems to have been ashore on the day, noted that such noises formed part of the usual celebrations of the monarch's birth day 'with great state and solemnity, and large bonfires,' all of which 'must have frightened' the natives.[2] For Southwell, the sound of celebration was comprehensible only to the colonists. The sound of ceremonial gunfire was a marker of possession and of shared meaning inaccessible to the Indigenous people, who, he supposed, could hear the *sounds* of celebration only as frightening *noise*.

Aboard HMS *Supply*, writing after the event however, Blackburn speculated on a different reception that might help to explain the unusually large gathering of Indigenous people seen close to the colony immediately after the celebration:

> Probably some of the Natives had been spectators from the woods. [They] Might take such a terrible noise as a denunciation of War and were collecting together for mutual assistance.

Blackwell's speculation was that the 'terrible noise' the Indigenous people heard betokened an organised and interactive response, namely war. In other words, the 'noise' of gunfire was not merely frightening, but was registered and understood within a shared, perhaps even universal language of warfare. For many Europeans in the eighteenth century, war was not merely organised violence, but an index of historical progress.[3] War was a practice whose variations changed as the societies waging war advanced in knowledge, manners and law – in short, as they became ever more 'civilised.' For many European contemporaries of Blackburn and Southwell, European warfare was the most advanced, the most technologically sophisticated, but also the most morally refined form of warfare. Civilised war embodied the unique accomplishment by which Europeans had, in Adam Ferguson's unforgettable phrase from his *History of Civil Society*, 'mingled politeness with the use of the sword.'[4] It was in this sense that Europeans were able to convince themselves that the 'terrible noise' of gunfire (and its dreadful effects) could be heard and understood by other peoples deemed less advanced as both an audible demonstration of the threat of war and as an audible demonstration of European civility, that is, of the norms that embodied their claims to a supposedly more advanced and more humane civilisation.

What Southwell and Blackburn heard on that June day is not in dispute, but the disparity in how they heard it encapsulates a deeper anxiety about the audible register of civility in cross-cultural encounters between 'Europeans' and 'Islanders' in the Pacific.[5] In this paper, I will explore the audible implications of these anxieties expressed in a selection of British accounts of such encounters from the late seventeenth to the late eighteenth century. I will argue that these anxieties were regularly assuaged not simply by recourse to the modulated civility of polite *sound*, but to the intrusive, disruptive and often terrifying recourse to discordant *noise*.[6] In arguing so, I want to complicate recent interpretations of European soundscapes that prioritise the cohesive force of purposive sound over the disruptive effects of intrusive noise. The audible context of early cross-cultural encounters in the Pacific demonstrates that mere *noise* – or what the European newcomers *heard* – played a crucial role in purposive activities (such as the establishment of imperial authority). Moreover, the sounds of early colonial encounters – what the European newcomers *listened to* – often did not confirm shared meanings and speech communities, but activated deep anxieties about their inability to communicate their civility to the Indigenous inhabitants they encountered. In this way, I want also to question the value of historical analyses that continue to separate normative and purposive 'sound' from valueless and random 'noise.'

Soundscapes of civility

A soundscape was famously defined by R. Murray Schafer as 'any acoustic field' consisting of 'events *heard* not objects *seen*.'[7] In Schafer's terms, the soundscape is a sonic environment that registers a physical space aurally, but does not necessarily make a normative distinction between events heard as sound or as noise. Such a normative distinction is implied, however, by historians who argue that modern industrial, mechanical and urban technologies enabled a '... transition from [a] relatively sound-scarce [premodern] social environment to the sound abundance of modernity,' a soundscape in which (from the mid-nineteenth century) ubiquitous and intrusive noise was perceived as 'an increasing urban blight … a condition of modern urban life rather than an episodic interruption of it.'[8] A condition of modernity, then, is that of 'listening *through*' the 'elemental' noises that screen purposive and integrative sound.[9]

Such a firm distinction between discordant noise and purposive sound, Emily Cockayne contends, was not always made in eighteenth-century discourse.[10] The term 'noise' may simply have been appended to any sound 'out of place.'[11] The place of sound or noise draws our attention back to the idea of a soundscape in which sounds were situated as purposive and integrative – as very much *in* place. The integrative force of premodern sound was the subject of Alain Corbin's analysis of the sounds of village bells. Their resonant sound was constitutive of the 'auditory space' of nineteenth-century French villages. This space was literally 'conjure[d] up' by the metre of reverberant sound.[12] The reverberation of sound, is, as Emily Thompson described it – 'the lingering over time of residual sound in a space.'[13] Reverberations move in resonant waves across and through physical space – they sweep across fields, or echo among cluttered city buildings, bouncing off walls as well as penetrating spatial subdivisions between inside and outside, public and private spaces. The reverberation of sounds was not just an effect of the thing making the sound, but of the urban architecture that enhanced or channelled the sound. Reverberant sound was also part of the mental 'architecture' of listeners for whom they denoted shared meanings – the announcement of celebration, the marking of time, the register of alarm.[14] Reverberating and resonant sounds linked the members of premodern societies in soundscapes that were at once integrative and purposive. They were integrative insofar as reverberant soundscapes flowed out across spaces inhabited by a wide range of listeners and hearers. Premodern soundscapes were supposed to be purposive in that the reverberations not only integrated those who heard within them, but also in that they were *listened to* and thus understood within shared frameworks of meaning.

This analysis sustains the pervasive assumption identified by Richard Cullen Rath, that the primacy of these soundscapes separates a supposedly primitive premodernity from the more civilised modern prioritisation of the ocular.[15] Integrative, reverberant soundscapes have come to be replaced, Thompson argues, by modern 'non-reverberant' sonic technologies in the twentieth century. Modern soundscapes have not become any less meaningful to those of us inhabiting them, but they have become more uniform, less specific to time and place, and thus amenable to its increasing commercialisation (and commodification by we moderns who do not only listen, but now consume sounds).[16] Lying behind this transition is the supposition that there was a premodern sonic sensibility characterised by the enmeshing of individuals by purposive sounds, typically in 'speech communities' that verbally express and represent shared meanings and identities, in contrast to a modern sensibility characterised either by the pervasion of increasing noise, or by the increasing sequestration or limiting of shared soundscapes.[17]

This idea of a premodern sonic sensibility is implied by the recent resurgence of interest in the discourse of civility in Early Modern Europe.[18] Concepts of civility arose in sixteenth-century Europe and emphasised the value of a 'civil' urban community with settled laws administered by legitimate political authorities enabling a 'sufficient' and prosperous life based on the inculcation of polite and courtly manners.[19] In the mass of instruction manuals and treatises on polite conversation, comportment, manners or rhetoric, civility emerged as a practical, performative, but above all, sonic virtue. Civility resonated in conversation, in public discourse, in studied oratory, in elegant music and in the manifold resonances of an industrious, prosperous and peaceful civil society.[20] By means of civility, bloody civil war could be averted, law, government and property would be secured, and Europeans could come to see themselves as embodying a common aspiration towards civilised societies marked by rapidly developing commercial economies, stable and secure sovereign states with imperial aspirations. Civility went hand in hand with mastery and empire over those deemed less civil, or in want of the supposed virtues of European civility.[21] Civility thus became entrenched by the beginning of the eighteenth century within schemes of historical progress or development that represented the states and societies of Western Europe, and their rapidly expanding empires, as the epitome of civilisation.

By the eighteenth century, Markku Peltonen contends, concepts of civility had become a European 'commonplace' expressing norms of 'polite conversation' endorsed across almost all political and religious divides.[22] This tells us very little about how Europeans across the broad sweep of the eighteenth-century thought that their civility could be communicated across divides that attenuated the possibilities for purposive and polite conversation, and led them instead to a reliance on mere noise. British accounts of colonial encounters in this period regularly belied worries about how the sonic communication of civility was heard by Indigenous people. They also betrayed anxieties about what they themselves heard when they listened to the Indigenous peoples they regularly presumed to be uncivilised. In considering the audible implications of this European anxiety, I make no claim for the priority of the aural over any other senses. Rather, I want to reflect on how, by listening to some of these encounters, we can hear again the ways that Europeans tried to convince *themselves* of their own civility by recourse not simply to civilised sound, but repeatedly to the brutality and banality of mere noise.

Hints, drums and a soft air

Concerns about how to make oneself heard and understood by 'the natives' suffused the 1768 'Hints' proffered to Captain James Cook by Lord Morton, President of the Royal Society, before Cook's first voyage to the Pacific aboard the *Endeavour*.[23] Morton's 13 pages of handwritten notes were intended as guidelines for Cook in his dealings with the presumed native inhabitants he was bound to encounter. The tenor of the 'Hints' is in many ways admirable. They advised Cook to prevent his sailors from committing outrages, to stop any violence against the Indigenous peoples, even if provoked, and maintained that Europeans could only claim land from them with their 'voluntary consent,' never by conquest.

Yet as Morton saw it, Cook and his men were the agents of a far superior civilisation, and for this reason, the burden lay upon the Europeans in particular to communicate their civil, civilised and civilising superiority. 'There are many ways' Morton wrote 'to convince them [the natives] of the superiority of Europeans, without slaying any of those poor people.' His account was based on a foundational assumption that the Europeans could make themselves understood by visual signs and audible cues that were unmistakable. 'Amicable signs may be made' he wrote, 'which they could not possibly mistake such as holding up a jug, turning it bottom upwards, to shew them it was empty, then applying it to the lips in the attitude of drinking' or 'Opening the mouth wide, putting the fingers towards it, and then making the motion of chewing, would sufficiently demonstrate a want to food.' Elsewhere he was concerned that the noises of European superiority might distress the natives – 'They should not at first be alarmed with the report of Guns, Drums, or even a trumpet.' 'But,' he helpfully suggested, 'if there are other instruments of music on board they [the natives] should be first entertained near the shore with a soft Air.'

In Morton's 'Hints,' European civility entailed a superiority that was to be both visually and audibly communicated. For all the stipulations about treating the 'natives' with civility and respect, even obtaining their consent, Morton's 'Hints' took for granted that Cook was the bearer of a superior civilisation and a legitimate British title to empire.[24] This legitimate claim to empire could be asserted in multiple ways – by dry legal assertion, by ceremonies of possession, by the unfurling of flags, by the utterance of commissions, by setting up 'suitable marks and inscriptions,' by saluting the King and by cheering the flag, many of which Cook was to do.[25] In each of these acts lay a pairing of sound and symbol, a visual and a sonic affirmation of European civility and civilisation in the orderliness, the sanctity, the unison and the presumed perpetuity of each act.

But for all Morton's certainty, the actual encounters between Europeans and Indigenous people in the Pacific, both before and after Cook's

1769 voyage, rarely presented opportunities for this kind of affirmation.[26] Rather, the records speak of the mystifying resistance of Indigenous people to accepting the superiority of civilised European claims to empire, and a persistent European anxiety that alternate means be found on which to rest their claims. Indeed, it would appear that the 'Hints' proffered to Cook about avoiding alarming drums and gunfire in preference to playing a 'soft Air' were inspired by the experiences of William Dampier almost 100 years before Lord Morton set pen to paper.

Dampier was the first Englishman to write about his interactions with the Indigenous peoples of Australia. His first account, the justly famous *New Voyage Round the World* of 1697, was the probable source for Johnathan Swift's unflattering depiction of the savage 'Yahoos' in *Gulliver's Travels*.[27] For this reason, it has tended to overshadow Dampier's later *Voyage to New Holland* of 1703. Here he recorded an effective episode of noisy Indigenous contempt to a desperate European attempt to display sonic superiority. On 31 August 1699, after several days of unavailing searches for fresh water, Dampier recorded that '9 or 10 of the Natives' confronted him and his shore party 'menacing and threatening of us, and making a great Noise.'[28] His efforts to make signs of peace being rebuffed, Dampier decided his best tactic was to try and surprise and 'catch one of them, if I could, of whom I might learn where they got their fresh water.' While making the attempt to cut off a group of them, one of his young men became closely engaged in a fight. Dampier came to his aid but narrowly missed being speared himself, whereupon, he wrote:

> I discharg'd my Gun to scare them, but avoided shooting any of them; till finding the young man in great danger from them, and myself in some; and that tho' the Gun had frighted them at first, yet they had soon learnt to despise it, tossing up their Hands, and crying *Pooh, Pooh, Pooh*; and coming on afresh with a great Noise, I thought it high time to charge again, and shoot one of them, which I did.[29]

This story of noisy confrontation could be said to have inaugurated a tradition of British colonial encounter in which the supposedly civilised were so confounded by Indigenous failure to recognise European superiority, that it prompted resort to the baldest, the noisiest and most fatal means of assertion. In doing so, Europeans betrayed a real anxiety that their presumed superiority was unsettled by the all-too-apparent failure of Indigenous people to play their part by acknowledging it. Of course, some Europeans asked why the Indigenous peoples they encountered did not acknowledge it. The unsettling possibility that such a question entertained was that the Indigenous peoples might perhaps possess their own forms of civility, and that there was nothing inherently superior about European civility and civilisation at all.

Pacific noises

Captain George Shelvocke seemed to make just this admission in his published account of the privateering journey he undertook in command of the *Speedwell* between 1719 and 1722. Shelvocke's aim was to raid Spanish shipping and colonies along the Pacific coast of the Americas. He reflected on the Indigenous inhabitants of the Baja Peninsula that '... there is a wide difference between what one would, upon the first sight, expect to find from them, and what they really are ...' for they are '... endued with all the humanity imaginable, and may make some nations (who would give these poor people the epithet of *Savages* and *Barbarians*) blush to think that they deserve that appellation more than they ....'[30]

Shelvocke in fact describes the Indigenous people he encountered in California and Chile as exemplifying an Ovidian 'golden age' where laws, government and property and all the artificial wants and superficial refinements of more civilised societies are unknown.[31] The contrast that informs his narrative, however, is not one between Pacific primitivism and European civilisation, but between the apparently natural simplicity of the Indigenous peoples, and the cruelty (and religious bigotry) of their supposedly civilised Spanish masters.[32] Shelvocke contrasts this Spanish cruelty with his own account of the meticulous civility and kindness with which he and his English crew engaged *and contracted* with the Indigenous Californians. In one case, Indigenous labour was encouraged, Shelvocke wrote, by having his crew set up a boiler on shore so that the Indians could be supplied with ample quantities of 'continually boiling pudding.'[33] And yet, notwithstanding the apparent amity of relations between the two groups, on his eventual departure, Shelvocke still thought it necessary to 'appear as awful to them as possible' and thus ordered 'five guns to be fir'd ... the noise of which mightily frightened them ....'[34]

Shelvocke's account of the voyage, and the interlude in California, was acrimoniously disputed and parodied by his Captain of marines, William Betagh, who claimed that the notion of Californian civility was a ridiculous fiction invented by Shelvocke to mask his own incompetence. 'There's none can forebear laughing' Betagh wrote, 'who reads his daily manner of feasting these *Californian* gentry with vast quantities of hasty-pudding: or his grave way of telling how respectfully and ceremoniously they returned his civilities ....'[35] According to Betagh, the Californian's 'wild ranting and frolicking' was indicative only of their savagery – for him they displayed neither acquiescent civility on the *Speedwell's* landing nor precipitate terror on their departure.[36]

In belittling Shelvocke as a fantasist peddling misleadingly 'romantick' descriptions of his 'Adventures,' Betagh paid scant regard to the descriptive scope of Shelvocke's prose. In one memorable passage, Shelvocke wrote, 'everything that one sees or hears in this place [the Western

American coast] is perfectly romantick ... [and] has a certain savage irregular beauty, which is not to be expressed ....' What struck him particularly, was the awesome sublimeness of the noise of nature that he evocatively described,

> ... the fall of waters, which one hears all around, would be agreeable to none but those who would indulge themselves, for a time, in a pensive melancholy ... nothing can be conceiv'd more dismally solemn, than to hear the silence of the still night destroy'd by the surf of the sea beating on the shore, together with the violent roaring of the sea-lions repeated all around by the echoes of deep vallies, the incessant howling of the seals, (who according to their age, make a hoarser or a shriller noise) so that in this confused medley, a man might imagine that he heard the different tones of all the species of animals upon earth mix'd together. ... All these, or any of these frightful noises would be sufficient to prevent the repose of any who had not been for some time enur'd to it.[37]

Was this remarkable passage the product of the 'scorbutic nostalgia' often felt by mariners on long journeys in the age of sail?[38] A more likely explanation is that it reflects Addison's account of the 'pleasures of the imagination' derived from 'the largeness of a whole view' of nature:

> Such are the Prospects of an open Champian Country, a vast uncultivated Desart, of huge Heaps of Mountains, high Rocks and Precipices, or a wide Expanse of Waters, where we are not struck with the Novelty or Beauty of the Sight, but with that rude kind of Magnificence which appears in many of these stupendous Works of Nature.[39]

For Addison, however, the pleasure inspired by such '*new* or *uncommon*' scenes could only be derived by the visual sense of grandeur and movement, 'where the Scene is perpetually shifting, and entertaining the Sight every Moment with something that is new.' For Shelvocke however, 'romantick' nature was not only seen but heard, and the sensation that audible change and movement inspired was decidedly melancholy.[40]

Shelvocke's reflections seem to have belied a sense that the Pacific was a realm of untamed nature in which European civilisation had but a tenuous hold. It was almost as if the natural world of the Pacific was so filled with noise as to drown out the modulated tones of civility. A similar notion of the uncivil noise of nature pervaded the journal of George Robertson, the Master aboard *HMS Dolphin* despatched by the Royal Navy to the Pacific in 1767. Throughout his narrative, Robertson likened the noise of nature with the noises made by the Indigenous peoples he and the crew encountered. When the *Dolphin* skirted along the Atlantic coast of Patagonia, for example, Robertson frequently wrote that the Patagonians '... keept hallowing and making a great Noise ...' at the ship such that most of the crew, long fed on stories of fantastic Amazonians and monstrous savages, supposed that '... their Voices was more like Bulls Bellowing or Lyons Roaring nor [than] the Common size of Men hallowing or talking.'[41] Throughout the journal, 'noise' was applied to the threatening utterances of beasts such as (dogs or wolves), or the mighty roaring of storms that threatened danger to the ship.[42] It was as if European civility required an announcement or a necessary precursor; an audible demonstration that registered above the sonic profile of breaking waves, hurtling storms or wild hallooing.[43] For Robertson as indeed for Shelvocke, that precursor was provided by cannon fire. Indeed, one of the first exchanges between the *Dolphin* and the Tahitians was punctuated by 'the Noice of the Gun,' which not only 'seemd to startle the Indians ...' but 'terefied' them. This seems to have been the intent.[44]

Whether through fear or friendship, however, trade with the Islanders was an absolute priority for the European newcomers. In common with Dampier and Shelvocke, Robertson's account demonstrates the ubiquity of trade as the medium of cross-cultural contact. Trade served a very practical purpose for all three, for each was in desperate need of supplies of fresh water, meat, but especially fruits and vegetables. By the time the *Dolphin* reached Tahiti, the vast majority of its crew and almost all the officers except Robertson were badly affected by scurvy. The Europeans comically signed and sounded for trade by imitating the sounds of hogs and chickens, while madly pointing to the shore, and Islanders reciprocated.[45] What the Tahitians quickly learned to value was anything made of iron or steel – nails, stanchions, hooks and knives. Although the terms of trade were quickly arranged, the trade was quickly undermined by the rapidly developing awareness that the crew were able to trade iron implements for sex ashore. So quickly did this trade develop that even those who had previously been so sick with scurvy as to be unable to get out of their hammocks crawled up on deck in the hope of being assigned to a shore party. Before long, every spare nail and stanchion had been pulled from the hull, and so many hammock hooks had been extracted by the eager sailors that all soon slept on the bare deck.

The manifold ways in which erotic dance and sexual encounter were understood by Tahitians as ways of 'galvanising the power of the gods to enhance the fertility of plants, animals and people,' have been well explored elsewhere.[46] For present purposes, it would seem that Europeans had little or no understanding that their material and sexual exchanges with the Islanders obliged them to their Tahitian partners, enmeshing them in powerful and perpetual spiritual and kinship relations. For the Europeans, all trade was 'traffick,' that is an exchange with definite temporal boundaries where one 'thing' (an object or service) is offered, paid for and provided, marking the trade as a momentary act with a beginning and an end. This profound misunderstanding had a number of consequences, one of the most important of which was the transmission of venereal disease to the island.[47] Other consequences of the commercialisation of sexual relations had a profound effect on the nature of cross-cultural communication. First, sexual commerce quickly devalued the currency used for the trade for food, causing massive inflation of prices for food. A second consequence of European misunderstanding of Tahitian exchange was to misread the audible and visual reprimands, registered in Robertson's words as 'loud cries' or 'hoot[ing] us,' as displays of uncivil hostility to which there was only one response, for as Robertson also described the dilemma: 'I thought myself under a necessity of using violent means ....'[48] But as he also wrote in a telling and repeated phrase that gunfire, even when lethal, 'hade not the desired Effect' for 'it only startled them a little' and they soon overcame their astonishment.

Robertson's account of the resort to gunfire powerfully suggests that it was motivated by the European's sensing that they were losing control of a volatile relationship. Hence, the trading encounters on the beaches and on board to *Dolphin* were peppered with powder and shot. As Robertson also testified, the reiterated punctuation of the cannon rarely had the intended effect.[49] This audible demonstration of civil superiority was prone to its own deflationary spiral, the more it was used the less terrifying was its effect, and the derision it evoked among the Tahitians inclined the British to more lethal demonstrations, poisoning the relationship on which they depended. Hence it was that Lord Morton's 'Hints' counselled pacific means to Captain Cook. When Cook entered the Pacific aboard the *Endeavour* only a year after Robertson's voyage on the *Dolphin*, he and his companions nonetheless soon found themselves confronted by the same dilemma, this time prompted not by the Tahitians but by the Maori.

Of guns and Hakas

When Cook and his crew arrived at New Zealand in late 1769, they encountered a people who in many ways baffled and confounded them. In the early encounters described in Cook's own journal, evident Maori hostility, but eagerness to trade jostled with the European's own erratic attempts to win or command their confidence by the noise of cannon, as well as the shouted reassurances that the newcomers would not 'eat them.'[50] One of the illustrators aboard the *Endeavour*, Sydney Parkinson, provided a series of telling visual and written descriptions of just how noisy and bustling the early European interactions with the Maori were. Among these was a curious image (Figure 10.1) entitled 'New Zealand War Canoe Bidding Defiance to the Ship' (by which he

Figure 10.1 Sydney Parkinson, 'New Zealand War Canoe Bidding Defiance to the Ship,' 1770, from 'A Collection of Drawings made in the Countries visited by Captain Cook in his First Voyage,' Add ms 23920 f050r, pen and wash drawing, 29.9 × 48.3 cm, British Library, London, UK/Bridgeman Images.

meant the *Endeavour*). This appears to be an incomprehensible image. The crew aboard the boat are depicted in contorted postures, arms and legs randomly splayed, each person crazily thrusting their limbs up or out; weapons, staffs and paddles are held or thrust at apparently random angles, their mouths agape and their tongues displayed, they stare about in all directions. Compounding this impression of disorganised contortion and random movement is the visual representation of the cacophony of noise we can imagine emanating from the canoe. Paddles are variously struck on its sides or thrust in the air, and warriors utter cries as they wildly gesticulate, stick out their tongues and twist or writhe about with apparently savage abandon. There is no order depicted in this Maori canoe; there is no unison either.

The puzzling thing about this image is that it seems to contrast so markedly with similar illustrations by both Parkinson (Figure 10.2) and one of the other illustrators on the voyage, Herman Spöring. In contrast to Spöring's distant and impersonal sketches of Maori canoes, Bernard Smith argues that Parkinson's figures are 'individualise[d] ...' and their 'dress and weapons' are pictured with 'clarity and precision,' giving 'each figure a set, almost statuesque position.' 'It is,' he wrote, 'a highly self-conscious composition.'[51] What Smith meant by this was that Parkinson was trying here for a deliberate series of effects. The first effect

Figure 10.2 Sydney Parkinson, 'New Zealand War Canoe,' 1770, from 'A Collection of Drawings made in the Countries visited by Captain Cook in his First Voyage,' Add ms 23920 f049r, pen and wash drawing, 29.8 × 48.2 cm, British Library, London, UK/Bridgeman Images.

was to capture an individuated neoclassical sensibility in the posing of the individual warriors (an effect perhaps better displayed in some of Parkinson's individual studies). Another effect altogether was the deliberate attempt to portray the group humorously.[52] Here, Smith made much of Parkinson's description of the Maori 'Heivos' (or Hakas) as affording him and the crew the demonstration of 'a truly comic act.'[53]

The Maoris themselves did not believe they were providing a demonstration of comedy. Far from it! The 'true' comedy lay, Smith argues, in Parkinson's perspective on the group as a naive and maladroit dance. Here, Smith relied on the evidence that Parkinson carried with him aboard the *Endeavour* a copy of William Hogarth's *Analysis of Beauty*. Hogarth's *Analysis* employed the depiction of 'Country Dancing' (a painting he had undertaken in 1745) to illustrate the principles by which the flowing, 'waving,' graceful elegance of movement could be represented. The country dancers, Hogarth wrote, 'make a delightful play upon the eye, especially when the whole figure is seen at one view.'[54] Their movement was to be represented 'in a composed variety of lines, chiefly serpentine' in an elegant sinuous waving line, whereas, he continued 'the dances of barbarians are always represented without these movements, being only composed of wild skipping, jumping, turning round, or running backward and forward, with convulsive shrugs, and distorted gestures.'[55] Significantly, Smith quotes these same lines from

Hogarth to make the point that the 'reflexive and interacting character of the situation' that Parkinson witnessed from the side of the *Endeavour* was inherently comical demanding the 'brusque angular lines' that both he and Hogarth had considered typical of 'barbarians.'[56]

The problem here is that Parkinson's image of the Maori war canoe does not conform to his own written descriptions of what he saw and heard. Parkinson's journal from 15 October 1769 (the one quoted by Smith to suggest that Parkinson's image was intended to be comical) indicates that European interactions with the Maori were noisy and bustling:

> In the morning [Parkinson wrote] ... nine canoes came to us, in which were one hundred and sixty of the natives ... [who were] shaking their lances and bone bludgeons at us, talking very loud and blustering, lolling out their tongues, and making other signs of defiance. We did all we could to make them peaceable, but to no purpose, for they seemed, at length, resolved to do us some mischief; coming along-side of the ship again, and threatening us, we fired one of our guns ... over their heads: they looked upon us for some time with astonishment, and then hastened away as fast as they could. ... They gave us two Heivos [hakas], in their canoes ... They beat time with their paddles, and ended all at once with the word Epaah; at the same instant striking their paddles on the thwarts [of the canoes] ...[57]

Parkinson here clearly recognised that the Heivo or Haka was performed not as a chaotic jumble (Figure 10.1) but in regulated time, to the rhythm of simultaneous striking of paddles upon the thwarts of the canoe, and to the orchestrated ending all at once. Joseph Banks also recognised the metre and rhythm of the performance carried out with 'as much Justness as if the movers were animated by one Soul ... without a mistake [ever] being made.'[58]

Parkinson and Banks were certainly impressed by the order and regularity of the Maori Hakas. They were also struck by noises and sounds of the Maori. Consider this entry from Parkinson's journal from 31 October 1769:

> ... seven canoes came off to us in a hostile manner, brandishing their lances, and waving their paddles. One of these canoes was very large, and had between fifty and sixty people in her; some of them gave us an heivo; and one of them, a priest, as we supposed, talked very much. They kept paddling about us, calling out to us ... [that] if we would go on shore they would beat us with their patta pattoos; and, being apprehensive that if we suffered them to approach nearer to us, we might be obliged to offer violence to them, the captain

> ordered a gun ... to be fired over their heads, the report of which terrified them so much, that they paddled away till they had got, as they supposed, out of our reach, and then they stopped, and held a consultation; after which they seemed as if they intended to return, and we fired another gun ... and then they made as fast as possible to the shore.[59]

Here noise serves to register Maori martial courage and ferocity in issuing threats. Noise also registers Maori orderliness – in the authority of the priest figure – in the unity of their Heivo or Haka – and in the consultation that they carry out among themselves to concert a strategy. But above it all, noise registers potential violence – not just from the supposedly barbarous Maori, but equally from the purportedly civilised British! For the British of course, the threat of Maori violence was accompanied by the terror of their purported cannibalism – perhaps hinted at in Parkinson's depiction of a preserved head held aloft in a Maori canoe (Figure 10.2), its eyes making contact with the viewer's own. Europeans regarded the violence of warfare as an index of civilisation.[60] The Maori unsettled Europeans by their tenacious and orderly courage in war – an apparently civilised virtue – but their purported practice of cannibalism was regarded by Europeans as baffling to say the least.[61]

Here as elsewhere, the Maori represented a puzzle – baffling both in their evident similarity and apparent difference from Europeans. They might both, Maori and British, have failed to read one another's true intentions leaving only the brute noise of gunpowder as the last communicative resort, but Parkinson heard something else in the interactions of the Maori with the British that struck another familiar and unsettling chord:

> ... the natives came on board and trafficked with us ... They were very cunning in their traffic, and made use of much low artifice. ... The natives, both on board and on shore, behaved with great civility, and, at night, they began to heivo and dance in their manner, which was very uncouth; nothing could be more droll than to see old men with grey beards assuming every antic posture imaginable, rolling their eyes about, lolling out their tongues, and, in short, working themselves up to a sort of phrenzy.[62]

There is a neat juxtaposition here of two European shibboleths – polite but cunning civility on the one hand, and an uncouth if droll barbarism on the other. But Parkinson's comments seem to suggest that the Maori were able to combine both qualities, and thereby to unsettle European usage of either term. Notably, Maori 'civility' was displayed in the activity of traffick or trade, an activity that did not solely consist in polite and deferential exchange, but in truck and barter involving every 'low artifice.' The phrase is significant. Civility was a performative virtue, it consisted in the conduct of individuals schooled and trained in politeness and manners, but it was for all that a highly artificial virtue. Maori civility appears to have impressed Parkinson precisely because of the ease with which they took to what Europeans regarded as a chief accomplishment of civility – the activity of traffick – with as much enthusiasm as ingenuity.

Conclusion

In providing his 'Hints' for Captain Cook, Lord Morton had suggested that European superiority was easily demonstrable by visual sign and by audible sound – by playing a soft musical 'air,' the natives would be put at their ease. As Parkinson himself observed, however, notions of superior European civility and inferior Maori barbarism were unsettled by audible signs. The defiant Maori Haka seemed to testify to their savage fierceness, but the conversational interaction of traffick registered their civility. Indeed, he was afraid that the Maori were so insensible of their presumed inferiority that it left Europeans only one recourse to demonstrate their supposed superiority – by the noise and dreadful effects of cannon fire.

It would be tempting to see Parkinson's images as an earnest and Enlightened effort to render the humanity of Maori for a European audience, to capture the repellent and baffling about them, as well as the admirable and universal. Whatever the merit in that view, I think this image should also be understood allegorically. By means of the symbolic representation of the discordant and disjointed noise of the Haka, Parkinson allegorised the Maori as primitively deficient in civility.[63] In perhaps the most celebrated allegory of Enlightened self-imaging, Gianbattista Tiepolo's 'Apollo and the Continents' (1753), the figure of Europe is pictured as the epitome of civilisation. Richly dressed, she wears a crown, and sits regally and still upon her throne surrounded by fine neoclassical architectural structures. The gathering of attendants paying court to Europe is redolent of European science, religion, arts and knowledge. Positioned prominently among them, a musical ensemble plays – a veritable performance of the European sophistication that Lord Morton recommended to Captain Cook. But prime position is given not to the musicians, but to a uniformed officer sitting upon the barrel of an enormous cannon.[64]

The cannon was a symbol to European contemporaries of their apparent mastery of science. Cannon fire came to be regarded as a superlative expression of European humanity in the Enlightenment period.[65] For many Europeans, artillery embodied an apparent mastery of science. It symbolised science at the disposal of humanity – a machine that could be engineered to deliver shattering power, but a machine whose effects (as both David Hume and Voltaire averred) drove ever-finer calculations about how to spare its use. And thus paradoxically, brutal firepower came to symbolise European civility.

Behind Tiepolo's rich allegorical imagery lay the bold conceit that the register of European civility echoed in gunfire. This is how cannon fire operated in Shelvocke's, Robertson's, Cook's and Parkinson's journals. It was an audible demonstration of the superiority of European civility. But even in its lamentable lethality, the cannon was also a demonstration of European 'humanity.' Firing the cannon demonstrated European ingenuity and science, the fruits of European language and labour, while also supposedly showing the European's concern to limit 'savage' violence by instilling fear. If the European Enlightenment was imbued with the idea of a universal humanity, it was a humanity still heavily laden with presumptions of superior civility that, in the absence of obvious Indigenous acquiescence to it, required demonstrable, audible affirmation as much in civil sound as civil noise.

Notes

1 David Blackburn to Richard Knight, 12 July 1788, David Blackburn Letters and Papers, Mitchell Library, ML Safe 1/120.

2 Daniel Southwell to Mrs Southwell, 12 July 1788, in *Journal and Letters of Daniel Southwell*, *Historical Records of New South Wales*, Vol. 2, ed. F.M. Bladen (Sydney: Government Printer, 1893), Appendix D, 668–732.

3 Bruce Buchan, 'Pandours, Partisans and *Petite Guerre*: Two Dimensions of Enlightenment Discourse on War,' *Intellectual History Review* 23, no. 3 (2013): 329–47. On sound and war see: Neil Ramsey, 'Listening to War: Sound and Noise in Romantic Era Military Writing,' *Republics of Letters* 5, no. 2 (2017): 1–12.

4 Adam Ferguson, *An Essay on the History of Civil Society* (1767; Edinburgh: Edinburgh University Press, 1966), 199.

5 I follow Thomas in the use of this terminology: Nicholas Thomas, 'The Age of Empires in the Pacific,' in *Pacific Histories: Ocean, Land, People*, eds. David Armitage and Alison Bashford (Basingstoke: Palgrave, 2014), 79.

6 Hillel Schwartz defines 'noise' as that which 'disrupts the initial stillness' and is 'forever conditioned by nostalgic yearnings for a calm that is no more, pleadings for a peace to be restored, oracles of a harmony yet to come.' Schwartz, *Making Noise: From Babel to the Big Bang and Beyond* (New York: Zone Books, 2011), 20.

7 R. Murray Schafer, *The Soundscape: Our Sonic Environment and the Tuning of the World* (1977; Rochester, NY: Destiny Books, 1994), 7–8.

8 David Goodman, 'Distracted Listening: On Not Making Sound Choices in the 1930s,' in *Sound in the Age of Mechanical Reproduction*, eds. David Suisman and Susan Strasser (Philadelphia: University of Pennsylvania Press, 2010), 16.

9 Schwartz, *Making Noise*, 86–91. See also David Ellison, 'The Call Came from Inside the House: Supernatural Noise and Domestic Incivility,' *Republics of Letters* 5, no. 2 (2017): 1–12.

10 Emily Cockayne, *Hubbub. Filth, Noise, and Stench in England, 1600–1770* (New Haven, CT: Yale University Press, 2007), 112–13.

11 On sounds 'out of place,' see Bruce Buchan, 'Civility at Sea: From Murmuring to Mutiny,' *Republics of Letters* 5, no. 2 (2017): 1–14.
12 Alain Corbin, *Village Bells: Sound and Meaning in the Nineteenth-Century French Countryside*, trans. Martin Thom (New York: Columbia University Press, 1998), 96.
13 Emily Ann Thompson, *The Soundscape of Modernity: Architectural Acoustics and the Culture of Listening in America, 1900–1933* (Cambridge, MA: MIT Press, 2004), 3.
14 Corbin, *Village Bells*, 191.
15 Richard C. Rath, *How Early America Sounded* (Ithaca, NY: Cornell University Press, 2003), 47–48.
16 Thompson, *Soundscape of Modernity*, 233.
17 Maarten Walraven, 'History and its Acoustic Context: Silence, Resonance, Echo and Where to Find them in the Archive,' *Journal of Sonic Studies* 4, no. 1 (2013), accessed 20 February 2014, http://journal.sonicstudies.org/vol04/nr01/a07.
18 Something similar is hinted at by Schwartz, *Making Noise*, 49.
19 Peter Burke, 'A Civil Tongue: Language and Politeness in Early Modern Europe,' in *Civil Histories: Essays Presented to Sir Keith Thomas*, eds. P. Burke, B. Harrison and P. Slack (Oxford: Oxford University Press, 2000), 34–38; Cathy Shrank, 'Civil Tongues: Language, Law and Reformation,' in *Early Modern Civil Discourses*, ed. Jennifer Richards (Basingstoke: Palgrave, 2003), 21–25.
20 Norbert Elias, *The Civilizing Process, Vol. 2: Power & Civility*, trans. E. Jephcott (1969; New York: Pantheon, 1969), 86–90. As Cockayne suggests, however, in the eighteenth century, 'noise' came to be associated by those in 'polite' society with the sounds emanating from the poorer classes. See Cockayne, *Hubbub*, 122.
21 John Hale, *The Civilization of Europe in the Renaissance* (London: Harper Collins, 1993), 360.
22 Markku Peltonen, 'Politeness and Whiggism, 1688–1732,' *The Historical Journal* 48, no. 2 (2005), 396–97.
23 An original copy of Lord Morton's 'Hints,' dated 10 August 1768, can be viewed online at the National Library of Australia's 'National Treasures' collection, accessed 10 July 2012, http://nationaltreasures.nla.gov.au/3E/Treasures/item/nla.ms-ms9-113-s003.
24 See, for example, Bruce Buchan, *The Empire of Political Thought, Indigenous Australians and the Language of Colonial Government* (London: Pickering and Chatto, 2008), 33–52. This legitimate title was confirmed in Cook's 'Instructions' that authorised him to take possession of parts of New Holland (Australia) 'with the consent of the natives.'
25 On the origins, uses and meanings of these rituals, see Patricia Seed, *Ceremonies of Possession in Europe's Conquest of the New World, 1492–1640* (Cambridge: Cambridge University Press, 1995).
26 Asserting sovereignty and exercising it were quite separate activities. As Chaplin observes, the European progress of empire in the Pacific proceeded 'belatedly and brokenly.' See Joyce Chaplin, 'The Pacific Before Empire, c. 1500–1800,' in *Pacific Histories*, 64.
27 William Dampier, *A New Voyage Round the World* (1697; London: Hummingbird Press, 1998).
28 William Dampier, *A Voyage to New Holland: The English Voyage of Discovery to the South Seas in 1699* (1727; Stroud: Nonsuch Press, 2006), 107.
29 Dampier, *A Voyage to New Holland*, 108.

30 George Shelvocke, *A Voyage Round the World by the way of the Great South Sea, Perform'd in the Years 1719, 20, 21, 22 in the Speedwell* (London: J. Senex et.al, 1727), 405–6.
31 Shelvocke, *Voyage Round the World*, 110.
32 Shelvocke, *Voyage Round the World*, 111.
33 Shelvocke, *Voyage Round the World*, 394, 395–97.
34 Shelvocke, *Voyage Round the World*, 398.
35 William Betagh, *A Voyage Round the World. Being an Account of a Remarkable Enterprize, Begun In the Year 1719, Chiefly to Cruise on the Spaniards in the Great South Ocean* (London: T. Combes, 1728), 216.
36 Betagh, *Voyage Round the World*, 217.
37 Shelvocke, *Voyage Round the World*, 257–58.
38 Jonathan Lamb, *Preserving the Self in the South Seas 1680–1840* (Chicago: University of Chicago Press, 2001), 123.
39 Joseph Addison, *The Spectator*, no. 412 (1712), in Joseph Addison and Richard Steele, *The Spectator*, ed. G. Gregory Smith, 4 vols. (London: J. M. Dent, 1945), 3:279–280. Thanks to Linda Andersson-Burnett for drawing my attention to this connection.
40 A precursor to later, more celebrated accounts of the 'sublime' generated by 'aweful' noise. See Schwartz, *Making Noise*, 147–48.
41 George Robertson, *The Discovery of Tahiti. The Journal of Second Voyage of H.M.S. Dolphin Round the World by George Robertson 1766–1768* (London: Hakluyt Society, 1948), 23, also 93.
42 Robertson, *Discovery of Tahiti*, 48, 51, 63–64.
43 Consider also Dampier's description of the noise of whales about the ship: '... we had in the Night Abundance of Whales about the Ship, some a-head, others a-stern, and some on each side blowing and making a very dismal Noise ... Indeed the noise that they made blowing and dashing the sea with their tails, making it all of a Breach and Foam, was very dreadful to us, like the breach of the Waves in very Shoal-water ....' See Dampier, *A Voyage to New Holland*, 101.
44 Robertson, *Discovery of Tahiti*, 126.
45 Robertson, *Discovery of Tahiti*, 137.
46 Anne Salmond, *Aphrodite's Island: The European Discovery of Tahiti* (Berkeley: University of California Press, 2009), 26.
47 Anne Salmond, *The Trial of the Cannibal Dog: Captain Cook in the South Seas* (London: Penguin, 2004), 49.
48 Robertson, *Discovery of Tahiti*, 144–45.
49 Robertson, *Discovery of Tahiti*, 148, 163.
50 James Cook, *Captain Cook's Journal During His First Voyage Round the World Made in H.M. Bark Endeavour 1768–71*, ed. W.J.L. Wharton (London: Elliot Stock, 1893), facsimile edition (Adelaide: Libraries Board of South Australia, 1968), 130–37.
51 Bernard Smith, *Imagining the Pacific: In the Wake of the Cook Voyages* (New Haven, CT: Yale University Press, 1992), 87.
52 This interpretation contrasts with that of Thomas, who detects the evident hostility in Parkinson's and Spöring's early drawings of Maori canoes. See Nicholas Thomas, *Discoveries: The Voyages of Captain Cook* (London: Penguin, 2003), 95.
53 Sydney Parkinson, *A Journal of a Voyage to the South Seas in His Majesty's Ship, The Endeavour*, ed. Stanfield Parkinson (London: Stanfield Parkinson, 1773), 93.
54 William Hogarth, *The Analysis of Beauty. Written with a View of Fixing the Fluctuating Ideas of Taste* (London: J. Reeves, 1753), 147 and 150.
55 Hogarth, *Analysis*, 150.

56 Smith, *Imagining the Pacific*, 89.
57 Parkinson, *Journal of a Voyage*, 92.
58 Joseph Banks, *The Endeavour Journal of Joseph Banks 1768–1771*, ed. J.C. Beaglehole, 2 vols. (Sydney: Angus and Robertson, 1962), 2:11.
59 Parkinson, *Journal of a Voyage*, 100.
60 John Gascoigne, *Captain Cook: Voyager Between Worlds* (London: Continuum, 2007), 115.
61 Thomas, *Discoveries*, 107.
62 Parkinson, *Journal of a Voyage*, 96. Parkinson also described the Maori as 'behaving very civil to us' on page 97.
63 Bruce Buchan, 'Images of Asia in European Enlightenment Political Thought, c.1600–1800,' in *Western Political Thought in Dialogue With Asia*, eds. C.J. Nederman and T. Shogimen (Lanham, MD: Lexington Books, 2008), 65–86.
64 The officer pictured here was the architect of the Residenz and colonel of artillery, Bathasar Neumann.
65 Bruce Buchan, 'Enlightenment Perspectives on War and Peace,' in *Visions of Peace: Asia and the West*, eds. T. Shogimen and V. Spencer (Farnham: Ashgate, 2014), 127–30.

Bibliography

Primary sources

Addison, Joseph and Richard Steele. *The Spectator* (1711–1714). Edited by G. Gregory Smith. 4 vols. London: J. M. Dent, 1945.

Banks, Joseph. *The Endeavour Journal of Joseph Banks 1768–1771*. Edited by J.C. Beaglehole. 2 vols. Sydney: Angus and Robertson, 1962.

Betagh, William. *A Voyage Round the World. Being an Account of a Remarkable Enterprize, Begun in the Year 1719, Chiefly to Cruise on the Spaniards in the Great South Ocean*. London: T. Combes, 1728.

Blackburn, David to Richard Knight, 12 July 1788. *David Blackburn Letters and Papers*. Mitchell Library, ML Safe 1/120.

Cook, James. *Captain Cook's Journal During His First Voyage Round the World Made in H.M. Bark Endeavour 1768–1771*. Edited by W.J.L. Wharton. London: Elliot Stock, 1893, facsimile edition, Adelaide: Libraries Board of South Australia, 1968.

Dampier, William. *A New Voyage Round the World* (1697). London: Hummingbird Press, 1998.

Dampier, William. *A Voyage to New Holland: The English Voyage of Discovery to the South Seas in 1699* (1729). Stroud: Nonsuch Press, 2006.

Ferguson, Adam. *An Essay on the History of Civil Society*. 1767. Edinburgh: Edinburgh University Press, 1966.

Hogarth, William. *The Analysis of Beauty. Written with a View of Fixing the Fluctuating Ideas of Taste*. London: J. Reeves, 1753.

Morton, Lord. 'Hints,' dated 10 August 1768. National Library of Australia. Accessed 10 July 2012. http://nationaltreasures.nla.gov.au/3E/Treasures/item/nla.ms-ms9-113-s003.

Parkinson, Sydney. *A Journal of a Voyage to the South Seas in His Majesty's Ship, The Endeavour*. Edited by Stanfield Parkinson. London: Stanfield Parkinson, 1773.

Robertson, George. *The Discovery of Tahiti. The Journal of Second Voyage of H.M.S. Dolphin Round the World by George Robertson 1766–1768.* London: Hakluyt Society, 1948.

Shelvocke, George. *A Voyage Round the World by the Way of the Great South Sea, Perform'd in the Years 1719, 20, 21, 22 in the Speedwell.* London: J. Senex et al., 1727.

Southwell, Daniel to Mrs Southwell, 12 July 1788, in *Journal and Letters of Daniel Southwell, Historical Records of New South Wales*, 2 vols, edited by F.M. Bladen, 668–732. Sydney: Government Printer, 1893.

Secondary sources

Buchan, Bruce. 'Civility at Sea: From Murmuring to Mutiny.' *Republics of Letters* 5, no. 2 (2017): 1–14.

Buchan, Bruce. *The Empire of Political Thought, Indigenous Australians and the Language of Colonial Government.* London: Pickering and Chatto, 2008.

Buchan, Bruce. 'Enlightenment Perspectives on War and Peace.' In *Visions of Peace: Asia and the West*, edited by T. Shogimen and V. Spencer, 127–30. Farnham: Ashgate, 2014.

Buchan, Bruce. 'Images of Asia in European Enlightenment Political Thought, c.1600–1800.' In *Western Political Thought in Dialogue with Asia*, edited by C.J. Nederman and T. Shogimen, 65–86. Lanham, MD: Lexington Books, 2008.

Buchan, Bruce. 'Pandours, Partisans and *Petite Guerre*: Two Dimensions of Enlightenment Discourse on War.' *Intellectual History Review* 23, no. 3 (2013): 329–47.

Burke, Peter. 'A Civil Tongue: Language and Politeness in Early Modern Europe.' In *Civil Histories: Essays Presented to Sir Keith Thomas*, edited by P. Burke, B. Harrison and P. Slack, 31–48. Oxford: Oxford University Press, 2000.

Chaplin, Joyce. 'The Pacific Before Empire, c. 1500–1800.' In *Pacific Histories: Ocean, Land, People*, edited by David Armitage and Alison Bashford, 53–74. Basingstoke: Palgrave, 2014.

Cockayne, Emily. *Hubbub. Filth, Noise, and Stench in England, 1600–1770.* New Haven, CT: Yale University Press, 2007.

Corbin, Alain. *Village Bells: Sound and Meaning in the Nineteenth-Century French Countryside.* Translated by Martin Thom. New York: Columbia University Press, 1998.

Elias, Norbert. *The Civilizing Process, Vol. 2: Power & Civility.* 1969. Translated by E. Jephcott. New York: Pantheon, 1982.

Ellison, David. 'The Call Came from Inside the House: Supernatural Noise and Domestic Incivility.' *Republics of Letters* 5, no. 2 (2017): 1–12.

Gascoigne, John. *Captain Cook: Voyager Between Worlds.* London: Continuum, 2007.

Goodman, David. 'Distracted Listening: On Not Making Sound Choices in the 1930s.' In *Sound in the Age of Mechanical Reproduction*, edited by D. Suisman and S. Strasser, 15–46. Philadelphia: University of Pennsylvania Press, 2010.

Hale, John R. *The Civilization of Europe in the Renaissance.* London: Harper Collins, 1993.

Lamb, Jonathan. *Preserving the Self in the South Seas 1680–1840*. Chicago: University of Chicago Press, 2001.

Peltonen, Markku. 'Politeness and Whiggism, 1688–1732.' *The Historical Journal* 48, no. 2 (2005): 391–414.

Ramsey, Neil. 'Listening to War: Sound and Noise in Romantic Era Military Writing,' *Republics of Letters* 5, no. 2 (2017): 1–12.

Rath, Richard C. *How Early America Sounded*. Ithaca, NY: Cornell University Press, 2003.

Salmond, Anne. *Aphrodite's Island: The European Discovery of Tahiti*. Berkeley: University of California Press, 2009.

Salmond, Anne. *The Trial of the Cannibal Dog: Captain Cook in the South Seas*. London: Penguin, 2004.

Schafer, R. Murray. *The Soundscape: Our Sonic Environment and the Tuning of the World*. 1977. Rochester, NY: Destiny Books, 1994.

Schwartz, Hillel. *Making Noise: From Babel to the Big Bang and Beyond*. New York: Zone Books, 2011.

Seed, Patricia. *Ceremonies of Possession in Europe's Conquest of the New World, 1492–1640*. Cambridge: Cambridge University Press, 1995.

Shrank, Cathy. 'Civil Tongues: Language, Law and Reformation.' In *Early Modern Civil Discourses*, edited by Jennifer Richards, 19–34. Basingstoke: Palgrave, 2003.

Smith, Bernard. *Imagining the Pacific: in the Wake of the Cook Voyages*. New Haven, CT: Yale University Press, 1992.

Thomas, Nicholas. 'The Age of Empires in the Pacific.' In *Pacific Histories: Ocean, Land, People*, edited by David Armitage and Alison Bashford, 75–96. Basingstoke: Palgrave, 2014.

Thomas, Nicholas. *Discoveries: The Voyages of Captain Cook*. London: Penguin, 2003.

Thompson, Emily Ann. *The Soundscape of Modernity: Architectural Acoustics and the Culture of Listening in America, 1900–1933*. Cambridge, MA: MIT Press, 2004.

Walraven, Maarten. 'History and its Acoustic Context: Silence, Resonance, Echo and Where to Find them in the Archive.' *Journal of Sonic Studies* 4, no. 1 (2013). Accessed 20 February 2014. http://journal.sonicstudies.org/vol04/nr01/a07.

11 The sounds of incivility

Insults and abuse in early Sydney

Penny Russell

In convict Sydney, insults and coarse language hurtled through the air. Amid the clash and clamour of forced labour and military discipline, the urgent whisper of gossip and the rising hum of commercial trade, words were wielded as weapons. Sometimes they slid harmlessly from a thick skin or an ear well accustomed to abuse. Sometimes they stuck like poisoned darts, or unleashed a furious, even violent response. Most hung on the air – peremptory, imperious, insolent, impudent or provoking – for a fleeting moment and then were lost forever. We can no longer hear these voiced insults, but from time to time insult led to dispute, and dispute ended up in a court of law. Transcripts of such cases, transmitted to the Colonial Office or reprinted in the press, capture the echoes of audible insult and transfer them into silent text. Reading those texts two centuries later, we may strain to hear reverberations of the emotion that weighted their utterance and stung sensitive auditors. By paying careful attention to the physical and emotional responses they elicited, we can glean a better understanding of how insults *sounded* at the time. Drawing upon scant surviving evidence of expression, intonation and response, this chapter attempts to restore some muted voices to the soundscape of the penal colony and to understand better both what people heard, and what they chose to hear.

Certain words carried a sting that some men, at least, could not ignore. Epithets such as 'rascal' or 'scoundrel,' for example, carried connotations of dishonour almost as grave as accusations of cowardice, especially to military men. But context always mattered. 'The mode of expressing words are often more insulting than the words themselves,' Richard Atkins, Judge Advocate, claimed in 1799, while himself bringing an action for defamation against a fellow colonist.[1] Many agreed: civil cases were crammed with testimony about the manner, tone and accompanying gestures that gave insults their intentional force; in criminal cases, too, the defence sometimes drew attention to the extreme provocation of insulting words combined with aggressive manner. Although social hierarchies, more than the language used, generally determined the outcome of cases, defendants, witnesses and prosecutors alike were keen to recreate the original sound of conflict and contest.

Their eagerness to do so gives us some of the best evidence we have on the sounds of colonial incivility.

Historians of sound almost routinely observe that the discipline of history is in thrall to the visual. Historians, suggests Peter Bailey, are products of a 'hypervisual modern world,' and since the advent of print have been 'seeing and looking with ever greater intensity and sophistication, while the other senses have been ignored.'[2] Mark Smith similarly links the pervasive visual orientation and metaphors of history – its relentless search for 'perspective' and 'focus' – to 'the nineteenth-century preoccupation with visuality, the rise of print culture, and the long shadows cast by these developments.'[3] Nevertheless, he reminds us, texts are not exclusively visual. Texts rich in sonic metaphors and sonic description should be considered not merely as 'indirect' sources on sound, but rather as 'direct' sources on how those sounds were heard, and the cultural significance that was attached to them.[4] The act of seeking 'sound' in texts thus becomes an act of *listening* for meaning – which is perhaps a better metaphor than most for the historian's craft.

The work of sound historians has done much to amplify the rich murmur of non-verbal acoustics from the past. The status of the spoken word as sound, however, remains more ambiguous. This may simply be because the transfer of sound to text has a different effect when the sound is verbal. A text that describes a non-verbal sound may evoke the referent, the noise and the meaning that has been attached to it. The majority of texts that record the spoken word simply translate speech into text. While Smith reminds us to listen for the sound of the human voice in texts, he emphasises the authorial voice rather than the aural impact of reported speech.[5] Alan Atkinson, a historian more interested in communities of speech than most, has something different again in mind when he insists that 'Voices are the only real medium of the soul, and hunting for voices is the historian's essential task.'[6] For Atkinson, the significant identity of his historical subjects resides 'in what they felt, thought and said over a period of time' – their 'voices' are heard in the 'telling of stories,' rather than the varied sounds that issued from lips and larynx.[7]

Several Australian historians have investigated links between voice, speech and power in colonial society. Joy Damousi argues that judgements about character and the self were closely connected with the voice: particularly in the later part of the nineteenth and the early twentieth century, elocution, purity of voice and pronunciation were powerful measures defining Australian culture and determining a character's integrity and entrée into respectable society.[8] The penal era has drawn the attention of historians interested in the speech patterns and vocabulary of convicts, and the relations of power and authority in which they were embedded. As Amanda Laugesen argues, 'Language reveals the attempt to create an ordered society, but also reveals a society where disorder always threatened.'[9] Catie Gilchrist has analysed attempts by penal authorities to maintain silence and civility as markers of order in convict society, and the challenge to power represented by the profane utterances of convicts, their 'abominable discourses' and 'horrible talk.'[10] The profane language of convicts horrified authorities, she argues, because they believed that 'immoral speech was contaminating and contagious.'[11] Yet the pious voice of the apparently reformed convict was scarcely more reassuring: polite deference, piety and quietness could so easily be a charade. 'The convict voice generated many moral and cultural anxieties because it could not be trusted.'[12] Colonial voices, in these accounts, feature not only as the tellers of stories, but also as markers of status, authority, resistance, respectability, profanity or corruption.

In this chapter, I add to these valuable discussions the consideration of another aspect of voice: the speech that wounded dignity, flicked at raw nerves or evoked angry responses. I 'listen in' on direct and indirect accounts of speech within the courtroom, written by press reporters from their verbatim transcripts, and designed at once to inform and entertain their readers with the drama of testimony and counter testimony. Such 'snippets of talk' have been 'screened at least twice,' as Jane Kamensky observes: 'through the screen of recollection, and through the filter of writing.'[13] In the courtroom itself, additional filters were supplied by the formalities of procedure, the dictates of relevance and the vested interests of those testifying. But by focussing attention precisely on such filters, acknowledging that what found its way into print was not the original conversation but a series of impressions about its impact, we may in fact come closer to an appreciation of how people in colonial Sydney – across differences of power and social position – heard each other, and how they interpreted what they heard. My aim is to open up what Veit Erlmann describes as a 'twofold process of "hearing another person listen",' in order to 'pick up all these sounds adrift, these echoes, reverberations, hums and murmurs outside or in between the carefully bounded precincts of orderly verbal communication.'[14] I am interested in what Kamensky terms 'the politics of speech' – in listening for 'the voices, the silences, and the clamor amid which people … made sense of their lives, day by day.'[15]

Words were never just words – their cultural meaning depended utterly on the social contexts, large and small, which affected both the ways they were spoken, and the ways they were heard. The records generated by the law, Kirsten McKenzie reminds us, 'cannot be read as straightforward reflections of how individuals thought about themselves and their injured honour.'[16] But words mattered, in law and in communities. In the effort to ensure justice, offensive speech was carefully reported verbatim, along with accompanying detail about the way words were spoken, and the responses they elicited – which in turn gives the historian some access to the shifting valence of particular words in communities of neighbourhood and of law. Such evidence can reveal, as Nigel Worden has argued, 'deep-felt and instinctive beliefs which were usually too obvious to contemporaries to openly express.'[17] As Arlette Farge so elegantly puts it, testimonies in court cases may be seen as

> fragments of ethics. Fragments of ethics, in the sense that the stream of words each person used to describe herself and the events reflects an ethos, an aesthetic, a style, an imagination, and the personal link that connected the individual to the community. ... These words trace intimate outlines and capture the thousands of ways each person has of communicating with the world.[18]

In what follows, I examine 'fragments of ethics' and fragments of speech from convict Sydney and situate them within the complex auditory dimensions of colonial space.

***

On 29 October 1795, some pigs belonging to John Boston broke into a quarter-acre garden, planted with wheat, the property of Captain Joseph Foveaux of the New South Wales (NSW) Corps. It was not the first time the wayward swine had thus trespassed, but for one at least it was certainly the last. The soldiers of the NSW Corps had no love for John Boston, one of the few free settlers in the tiny and still very new penal colony. A self-proclaimed Jacobin, Boston's politics offended them, while his desire to enrich himself through trade – his main motive for emigration – threatened their commercial monopoly and growing political power.[19] Boston's pigs, though straying in breach of regulations without nose rings or yoke, might yet have been returned with a stern warning – or perhaps, as had happened on previous occasions, after a beating to discourage future trespass. But this time, the Quartermaster, Thomas Laycock, when he came upon the wandering pigs, ordered a private to shoot one of them. What was more, he selected for execution a particularly fine and pregnant sow. Informed of the awful deed, Boston rushed to the spot to find his prize pig stone dead. Helpful bystanders pointed him in the direction the executioners had taken. Boston ran after them, exclaiming 'Who [or perhaps 'Where'] is the damned rascal that killed my sow?'[20]

The word 'rascal' held special resonance for military men. The officers of the NSW Corps were famously touchy on matters of honour – not least because guarding a penal colony was hardly the most honourable branch of the service. But honour was also a sensitive area to foot soldiers who had been recruited, often, from the lowliest echelons of society. Little more than their uniforms and the turn of fortune separated them from the convicts they guarded. To be called a 'rascal' stripped them of the illusion of dignity, and they could resent the insult with a vehemence that suggested they knew their own vulnerability. As the Court of Civil Jurisdiction acknowledged in reference to this case, 'the Military is a School of Honour, and ... no term save that which brands him with the want of courage, can be more harsh in a Soldiers Ears than that of Rascal.'

Perhaps, then, Laycock was leaping to the defence of the honour of the Corps when he instantly demanded to know *whom* Boston was calling a rascal. Or perhaps he was simply, and deliberately, stirring mischief. Boston was happy to oblige. 'The man that shot my sow,' he retorted. He would argue later before the Court that these 'words of Passion' had been 'uttered in the Moment of impression in the absence of the parties' – and therefore could not in all seriousness have been resented. But Laycock at once gave the insult a specific target by calling to William Faithfull, the private who had carried out the execution. Telling Faithfull that Boston had called him a rascal, he 'commanded him to thresh him as long as he could.' Further argument and a scuffle ensued, in the course of which Faithfull struck Boston on the back of the head with the weapon he had nearest to hand – his loaded musket. The two men struggled for possession of the gun until Laycock prudently took it away. The sounds of the scuffle echoed through the streets. Boston was 'exclaiming ... Murder, and calling for Assistance'; his wife, 'terrifyed and alarmed had run to the spot,' and now blended her cries with his; while Laycock with 'savage Brutality' urged a second soldier, William Eaddy, 'to thresh [Boston] as long as he had life, and push'd him for that purpose towards him, asking him if he would allow that fellow to call a Soldier a Rascal.' Overpowered by force and numbers, Boston broke away and escaped through the back door of a neighbour's garden.

We know of this exchange of heated words and blows because Boston subsequently sued Laycock, Faithfull, Eaddy and another soldier he claimed was involved, Neil Mackellar, before the Court of Civil Jurisdiction, demanding £500 in damages. There was much more to the case than a bad-tempered response to an act of law enforcement against swinish trespass: the parties concerned knew each other well, and a previous exchange between Laycock and Boston had left both, it seems, harbouring feelings of resentment, which resurfaced during the hearing of the case. Legal historian Bruce Kercher has identified this as the first time the civil court was 'used to test the rights and powers of the military,' thus becoming 'a forum for a struggle over ideology and economic power.'[21] Lisa Ford adds that the case was 'one of the most important decisions ever handed down in the colony of New South Wales,' because it addressed a dilemma central to the constitution of the British Empire – 'what sort of Peace could be transposed from the metropolis to its colonial peripheries?'[22] In this 'fragile convict-ridden outpost,' the military had a greater role in preserving the King's Peace than would have been allowed at home.[23] The soldiers who thrashed John Boston and killed his sow argued that they occupied the position of a constable in England, and that Boston had threatened the peace by his seditious talk, his disorderly pigs and his own riotous conduct.[24] From their perspective, Ford suggests, Boston 'deserved his beating ... because he threatened a hybrid martial order in a fragile imperial periphery.'[25] Boston, in turn, 'contended for a different sort of Peace in the colony': one in which civil power curbed the privileges of the military, curtailed their monopoly of commerce in the colony and allowed free settlers like himself to prosper by dealing freely with convict customers and creditors.[26] As Kercher puts it, he 'held the bourgeois view that freedom was allied to the right to make money.'[27] In Parsons' view, the trespass of Boston's pigs presented 'a timely opportunity for the military to punish an enemy of the King and a trading rival': thus the 'poor sow was a political martyr.'[28] Boston ultimately won his case, but the damages awarded him were pitifully small, and he suffered another attack from soldiers on his way home from court. Civil order had been reasserted, but the power of the NSW Corps in the colony continued to hold sway until it overreached by deposing the governor in the so-called 'Rum Rebellion' of 1808.

Political, legal, economic and imperial contexts are essential, then, if we are to understand the causes and ramifications of this apparently trivial event, and help to explain its survival in the historical record. But for a historian interested in the auditory landscape of the colony, and the aural impact of insult, the case also yields a wealth of interesting detail, firmly grounded in local and interpersonal context. Both prosecutor and defendants managed their own cases without the aid of lawyers, and unashamedly questioned witnesses in ways designed to bring forth a favourable version of events. While Boston strove to make the case that his indignation was natural and no reasonable person would have heard his words as provocation, defendants brought a string of witnesses to testify that both his language and his manner had been aggressively hostile, constituting a direct and intolerable verbal assault on the soldiers' honour. The witnesses, all too obviously reporting only what they had been allowed to hear or encouraged to repeat, invented violent and abrasive speeches, or massaged them into innocuous explosions of mild wrath. Though the noise of their testimony distorts and drowns out the original speech of that memorable day, it supplies in its place evidence arguably of greater value to the historian, for it shows their own awareness of the potential effects of speech, intonation and manner. By inventing accounts of dialogue intended to incriminate or exonerate, the witnesses showed their own sense of the social weight of language, of insult and of tone of voice.

Thus Nicholas Devine, as a witness for Boston, told the court that he had seen Boston and Laycock 'talking together with much warmth.'

> I heard Mr. Boston say it was a rascally business, of affair[.] Mr. Laycock said, Do you call a Soldier a Rascal – Mr. Boston said, it was a Rascally act.

Asked by Mackellar, 'Did you hear Mr. Boston say that all the Soldiers were thieves and rascals, and abused them very grossly?' Devine was tersely uncooperative. 'I did not,' he replied. Another witness was equally unsatisfactory to the defence. When Mackellar asked, 'Did you hear the Plaintiff make use of any opprobrious language calling the Soldiers thieves and Rascals?' Thomas Rumbold replied, 'I heard Mr. Boston make use of the word Rascal, but do not exactly know who he applied it to.' Boston's witnesses thus supported his own account of events – that the words he uttered were an expression of natural indignation in response to acute provocation. The injury done to him by the death of his prized sow was certainly exacerbated by his belief that the execution was a deliberate display of contempt for himself. But he was adamant that the word 'rascal' had expressed his anger at the event, and had not been directed contemptuously towards a specific individual, or to the military as a whole.

Witnesses called by the defendants told a different story. 'I saw Boston coming along raving,' Sergeant William Jamison declared:

> he asked of 2 Soldiers where is the damned Rascal who has shot my Pig; – I know not what they said – but heard Boston say, 'he would cut off his head by––' he then went towards Captain Foveaux's Garden – ... I saw him strike Faithfull at the end of Captain Foveaux's Garden – he retreated and Faithfull followed and struck him with the Musquet ... when I came up they had both hold of the Musquet. Boston struggling to get the Musquet from him, calling Faithfull in the meantime you damned Rascal.

Pressed by Boston on some of the more dubious aspects of his testimony, Jamison was unshakeable: 'I heard you loud enough talking to either one or the other – you was talking loud enough to be heard distinctly 100 yards; You damned Rascal, you have shot my Pig.' The next witness, James Biggs, repeated and embroidered the story:

> I saw the Plaintiff coming along in a riotous manner – we met – I heard him call to a Soldier; Hollo you Soldier, where is the damned Villain of a Rascal who has shot my Pig; I told him it was shot in Captain Foveaux's Garden – he went thither – Mr. Laycock was at the Gate – the Plaintiff asked him where was the damned Villain of a Rascal who shot my Pig; Laycock called Faithfull who came – Mr. Boston asked him if he was the damned Rascal who shot his Pig – Faithfull said he was, why do you not keep him out of the

> Garden – he called Faithfull a damned Blackguard Rascal. ... [and of the soldiers] said you are all a d–– parcel of Rascals together.

Not just a rascal, then, but a damned villain of a rascal, a Blackguard rascal – all in direct accusation, of Faithfull and of the corporate body of the military. In the defence's narrative, Boston was not in the grip of 'natural' indignation, he did not speak with 'warmth': instead, he was 'raving' and 'riotous' – furious, unhinged and bent on offence. It was he who forced the quarrel on the soldiers, not the other way round. And the evidence of that lay in the tone of his speech.

Later in the day, as the court was about to adjourn, Boston informed them that Biggs had been seated outside throughout Sergeant Jamison's testimony, and 'within hearing of every word the Witness Jamison said.' Questions of proximity and distance, of what could have been heard, or overheard, and what might have been missed, echoed through the court case, in continual reminder that the larger politics of the case were fought out via contested acts of speech. This tiny community was held together, and riven apart, by colonial voices.

***

Context mattered. Tone of voice mattered. So too did intent to offend. But the words that were used mattered too, and in recognising this, we can get a better understanding of how the words themselves sounded to their hearers. Certain words carried connotations peculiar to time and place, and the level of offence given by those words depended on broader community understandings of their significance and application. As Peter Moogk has observed, 'it is difficult to be entirely original in one's choice of epithets. Each culture has its distinctive and patterned range of pejoratives which provide insights into the mentality and values of a cultural group.'[29] For an insult to be effective and damaging – to strike the ear as a particularly ugly sound – it must identify a victim with something that the society regards as wrong or evil. Sociologist Ann-Carita Evaldsson goes further, suggesting that the 'language practice of insulting' not only reflects, but may indeed have a constitutive role in social organisation and gendered identities.[30] In the final part of this chapter, I want to leave John Boston and his pig in order to trace something of the history of the word 'rascal' he so rashly employed, a word that resounded through convict Sydney in the early decades of the nineteenth century, acquiring new complexity of tone with each reverberation.

While it is debatable how far Laycock and Faithfull were, in reality, outraged by Boston's implied attack on military honour, there can be little doubt that for a free settler to call a soldier a 'damned rascal' would have flicked a raw nerve. Although by the twentieth century it had acquired a benign ring, as one that might be applied affectionately to a mischievous schoolboy or a thieving puppy, in the late eighteenth and early nineteenth centuries, the epithet 'rascal' – and its companion word, 'scoundrel' – had a very different effect. We find one version of its more damning resonance in Jane Austen's *Sense and Sensibility*, in the scene where Willoughby, stricken by the rumour that Marianne Dashwood has been brought to the point of death through heartbreak at his betrayal, pours out his self-loathing contrition to her sister Elinor. In this climactic scene, Willoughby three times describes himself as a 'rascal,' and once as a 'scoundrel' – thus voicing his awareness of his own base conduct in toying with Marianne's feelings when he had neither the intention nor the freedom to marry her. The words leave no doubt of the depth of his self-abasement.[31]

But a 'rascal' did not only refer to sexual immorality among elites. Its sting lay in its class-laden association with a man without possessions, without respectability, with no stake in society and no regard for his own self-worth. Twenty years after John Boston's case, the word festered in the heart of an Irish convict labourer named Michael Hoolaghan. As a member of a gang working on the Liverpool Road in 1814, Hoolaghan was offended by gatekeeper Edward Mayne's refusal to allow convict women from the Parramatta Female Factory to pass through his turnpike to visit the men at their huts. One memorable day, he waylaid Mayne's wife, demanding to know 'what Mayne meant by calling them ragged rascals.' He added, 'Let him look out the next time I get drunk, for then I'll serve him.' Not long afterwards, true to his promise, he appeared at the turnpike at midnight, armed, abusive and accompanied by a mate. While Mayne struggled with his unwanted visitors, two men who were staying with him came to his assistance. In the ensuing fight, both were killed. Hoolaghan's drunken attempt to insist on his right to respect, his indignant refusal of the epithet of 'rascal' and all that it connoted of degradation and lack of character, ended in murder and, for himself, the gallows.[32]

In 1816, a man named John Bolton gave evidence on behalf of another who had been accused of stealing a pig. The accused was convicted despite his testimony, and at the close of the trial, the prosecutor, John Webb, approached Bolton while they were still within the courthouse and called him a 'false swearing rascal or scoundrel.' Two witnesses attested to the exchange, though neither could remember precisely which word had been used, suggesting that both carried similar import. Webb himself, also clearly aware of the offensiveness of the terms, recalled his own words differently, claiming that he had merely asked, 'How came you to take such a false oath?' If those were indeed his words, Bolton's intemperate response appears disproportionate. Allegedly, he first retorted that if Webb dared to say such a thing again he would knock him down – and then, filled with uncontrollable wrath, delivered the threatened blow anyway. Charged immediately for this violent act in contempt of court, at his trial the following day he pleaded in mitigation 'the impulse of a mind strongly agitated by the calumny.' The judge sentenced him to 12 months in gaol – but at the same time severely rebuked the prosecutor for casting 'so unjustifiable an aspersion.'[33]

These two cases were among the very few reported instances of 'rascally' insults during the first two decades of the nineteenth century in Sydney. The first colonial newspaper, the *Sydney Gazette*, came into existence in 1803, and the recent digitisation of it as well as later additions to the NSW press makes it possible to measure with some confidence the pervasiveness and shifting significance of words like 'rascal' and 'scoundrel' in text and speech. A keyword search for the two terms (used together or interchangeably) in these papers reveals an interesting trend. In the *Gazette*, the words appear only seven times between 1806 and 1815, and only 24 times in the following decade. But then something happens. Between 1826 and 1830, the words occur 143 times in the *Gazette* alone and almost 200 more in the other papers that had risen up alongside it. In the early 1830s, usage in the *Gazette* rose to 222, and in the second half of that decade to 292. Across the growing body of Sydney papers as a whole, we find the words 1,091 times in 1836–1840. After 1840, usage begins to taper off.[34]

The sudden proliferation of these words of abuse in the press during the 1830s is not necessarily a reflection of how often they were used as demeaning insults in the daily speech of the colony. They were, in part, simply the product of a thriving press, as more, and longer, newspapers were published, and with increased frequency. Their various editors were also locked in political and commercial rivalries that inclined them to comment on each other's reporting of local events, encouraging repetition of particular themes and events. In addition, many local columns were filled with extracts from English periodicals, including fiction and journalism replete with imagined conversations, colourful insults and pejorative terms. But these factors alone do not explain the sharp increase in usage. Often the words were used in contexts that directly reflect the growing complexity of urban society in a penal colony. Where convict, ex-convict and free men and women literally rubbed shoulders as they worked together in varied occupations and hierarchical relationships, questions of reputation and human dignity took on a deeply political cast.[35] There was, moreover, acute sensitivity to status and reputation in a colony beginning to feel, and express, the desire to throw off the yoke of convictism that was seen as holding back political freedoms.[36] In this context, the press was increasingly preoccupied with the question of insult, and the demeaning language used towards social inferiors.

Robert Shoemaker has argued that in London, public insult declined in significance over the early modern period. Legal cases by the late eighteenth century were less likely than in earlier times to turn on the specific words used and their capacity to destroy character. Increasingly litigants sought damages for direct financial loss, stemming from injury to commercial reputation – and increasingly they were required to show that the insult had indeed produced such detrimental effect. The mere loss of character in the eyes of one's neighbours, he suggests, was less likely to be seen as a matter for legal remedy.[37] Cases of insult in Sydney during the 1830s were similarly more likely to succeed if they could show commercial or professional detriment. In 1834, Edward Salomon complained to the Supreme Court of NSW in Banco that after giving evidence in a civil case the preceding day, he had heard himself stigmatised as a 'perjured Jew scoundrel.' The private apology that had since been tendered to him could not, he argued, remove the effect of 'such an insult inflicted on him in so public a situation as the door of the Court-house, surrounded as it was by persons of respectability.' He owed it to the public to complain, lest others be discouraged from giving honest testimony in similar cases. The Court refused the application, arguing that they were not obliged to make a ruling where there had been no intimidation or prejudice of justice. They recommended Salomon, if he thought the expressions slanderous or likely to 'injure him in his business,' to seek remedy by civil action.[38] There is no evidence that he did so. His complaint appears to have been an attempt to protect his personal character, rather than his business, against insult, and for this, the courts were of limited value.

When solicitor Thomas Dean Rowe brought an action against merchant Caleb Wilson for £1,000 for calling him 'a perjured rascal and a scoundrel,' the defendant's counsel argued that the damage to Rowe's character had been immaterial: he was unlikely to suffer from 'what had been said … at a moment of hilarity, merely in the hearing of a few government men.'[39] The judge advised the special jury assembled to hear the case that significant damages would only be warranted if they were satisfied that the words were spoken 'calmly and deliberately with an intent to injure the plaintiff' – if spoken 'in the heat of passion,' they might look on it in a different light. The jury, finding 'the words were spoken, but as mere common abuse,' asked whether they were still obliged to find for plaintiff. Yes, said the judge, if spoken deliberately 'and concerning a professional man.' The jury promptly awarded the plaintiff a single farthing in damages.[40] Such transactional considerations of actual damage done were a long way from the debate over the touchiness of military honour in John Boston's case.[41]

Nevertheless, in a penal colony, certain insulting words retained significant power not only to sting, but to harm social standing. Class and criminal background did not align neatly in this mixed and highly sensitive society, where well-born and highly skilled convicts might still be employed in a variety of clerical and professional occupations.[42] The stigma of convictism was not applied equally or automatically to all convicted criminals, and the common employment of such euphemisms as 'government men' in place of the more damningly unambiguous 'convict' has been hailed by some historians as evidence of the freedoms and incipient egalitarianism of a convict society. But reference to the degraded status of a convict, in past or present life, was a readily available weapon, and one frequently employed in the context of political rivalry. It was mainly for political reasons, as Kirsten McKenzie has shown, that the conservative *Sydney Herald* called out the allegedly dubious commercial practices of Sydney auctioneers while making veiled but unmistakable reference to the most prosperous auctioneer of the day, Samuel Lyons, as a 'thrice-convicted convict.' When Lyons sued, successfully, for libel, the *Herald* waxed indignant. The law of libel, they asserted, was intended as the 'vindicator of good fame' and the protection of character. It should not be made the 'shield of the scoundrel,' a weapon for those who had foregone all claim to character to 'annoy and rob respectable persons.'[43]

Convict workers blended invisibly into the labour force only until it became expedient to call them out. In this fluid world of name-calling, certain newspapers took it upon themselves to defend the dignity of the working man, including the convict, from imperious and abusive authorities and employers, whose own moral conduct in many cases could little bear inspection. Edward Hall's *Sydney Monitor*, in particular, carved this line. In 1831, the *Monitor* noted with disapproval the 'brutal demeanour' and language of constables at the Police Office towards people taken before the Sydney Bench. '*Here* scoundrel! Go there you rascal! is the language of our well-behaved constables,' they observed, and recommended that Rossi, the Police Magistrate, should set the example by adopting a 'mild and courteous manner.'[44] A year later, condemning in florid prose the crimes of a once wealthy 'rogue,' they declared they had no pity for such as he. Their pity was reserved for the 'convict still in bondage' – poorly fed, inadequately clothed and often oppressed. 'He is abused – called scoundrel, rascal, convict, thief; and if, under all his troubles, he is not patient and meek, he is punished.' So long as the editors could wield a pen, they declared, 'the oppressed portion of such class will never want a friend.'[45]

The *Monitor* made good on this promise early in 1839 when Rossi's successor as Police Magistrate, Henry Croasdaile Wilson, provoked a media frenzy by taking a high-handed tone with the convict servants of another man. In this, however, the paper was hardly alone. Wilson was already *persona non grata* for his treatment of his own police constables, and not long afterwards would be dismissed from his position.[46] His flash of temper and its effects therefore fell on fertile media ground. The incident involved some convict labourers who had been set to work by Sir Maurice O'Connell to improve the road that led from his Darlinghurst home to the waterside. Though O'Connell claimed the road as his private property, his neighbours used it 'freely, and at all times, in their pleasure.' Finding that the work was delayed and sometimes even endangered when the men suspended work to doff their hats to the gentry, O'Connell ordered them not to do so. The next passer-by, unfortunately, was Wilson, accompanied by some ladies. Perceiving this as a slight to himself and his companions, Wilson demanded that the men touch their hats. Most complied, but one man, Todd, would not. Wilson ordered his police attendants to take Todd into custody; Todd resisted, and 'made use of some opprobrious expressions.' Brought to court the next day by Wilson with the complaint that he had failed to doff his hat, Todd was sentenced by the Bench to 50 lashes – not for the original alleged offence, which they deemed justifiable in the circumstances, but for the language he had used towards Wilson after he was handcuffed. The *Monitor* waxed indignant, so too did all the major papers other than the *Herald*. Wilson, wrote the *Monitor*, had 'called this man a "d—d scoundrel"' – language that he was only too accustomed to employing 'to all his constables and all his servants, the moment they displease him.' Todd, they considered, had been justifiably incensed, and Wilson himself deserved reproof for using 'violent language unbecoming a magistrate or a gentleman to the man, whom he on his part causes to go to the torture for doing the same by him.' The epithet 'scoundrel,' applied to a convict, in this instance was heard, by the *Monitor* at least, as the language of tyranny.[47]

Reading the shifting language of insult in colonial Sydney demands attention to context, to hierarchy and to the shifting valence of particular words. It is, in many ways, an act of looking – scrutinising print for evidence of meaning. But in the case of insults, the meaning was always partly aural. Their recipients heard, only too clearly, the shrill or strident note, the scoffing mirth, the haughty contempt of authority, the deliberate goading of malice. We too must listen for those silenced voices of the past, to understand that the same word might be savagely damning or casually abusive, perhaps even amicably jocular, depending on the context and tone of its delivery. But to fully understand the fluid force of epithets like 'rascal' and 'scoundrel,' and the different ways they were heard and resented by soldiers, free men and convicts, we must also strain to listen with the ears of the past. The recorded speech we now read as text was, in the first instance, always heard – and the hearing was itself a part of history.

Notes

1 Atkins v Harris, NSW Court of Civil Jurisdiction Proceedings, State Records NSW 2/8150 (1799), accessed 25 May 2018, www.austlii.edu.au/cgi-bin/sinodisp/au/cases/nsw/NSWSupC/1799/4.html.

2 Peter Bailey, 'Breaking the Sound Barrier,' in *Hearing History: A Reader*, ed. Mark M. Smith (Athens: University of Georgia Press, 2004), 23.

3 Mark M. Smith, *Listening to Nineteenth-Century America* (Chapel Hill: University of North Carolina Press, 2001), 6.

4 Mark M. Smith, 'Introduction,' in *Hearing History: A Reader*, ed. Mark M. Smith (Athens: University of Georgia Press, 2004), xv.
5 Smith, 'Introduction,' in *Hearing History*, xv.
6 Alan Atkinson, 'Writing about Convicts: Our Escape from the One Big Gaol,' *Tasmanian Historical Studies* 6, no. 2 (1999): 25.
7 Atkinson, 'Writing about Convicts,' 26.
8 Joy Damousi, '"The Filthy American Twang": Elocution, the Advent of American "Talkies," and Australian Cultural Identity,' *American Historical Review* 111, no. 2 (2007): 397. See also Joy Damousi, *Colonial Voices: A Cultural History of English in Australia* (New York: Cambridge University Press, 2010).
9 Amanda Laugesen, *Convict Words: Language in Early Colonial Australia* (Melbourne: Oxford University Press, 2002), ix.
10 Catie Gilchrist, '"A Life of Noisy Riot, of Filth, Indecency and Profaneness": The Convict Voice and the Bourgeois Imagination,' *Journal of the Royal Australian Historical Society* 92, no. 1 (2006): 34.
11 Gilchrist, 'A Life of Noisy Riot,' 38.
12 Gilchrist, 'A Life of Noisy Riot,' 41.
13 Jane Kamensky, *Governing the Tongue: The Politics of Speech in Early New England* (New York: Oxford University Press, 1997), 10.
14 Veit Erlmann, 'But What of the Ethnographic Ear? Anthropology, Sound, and the Senses,' in *Hearing Cultures: Essays on Sound, Listening and Modernity*, ed. Veit Erlmann (Oxford: Berg, 2004), 18.
15 Kamensky, *Governing the Tongue*, 10. My focus in this chapter is on insults within white male colonial culture. For an examination of voice, gender and power, see Penny Russell, 'Throwing down the Gauntlet: Voice, Power and Sexual Violence in Penal New South Wales,' in *A Cultural History of Sound, Memory and the Senses*, ed. Joy Damousi and Paula Hamilton (London: Routledge, 2017), 93–108.
16 Kirsten McKenzie, 'Defining and Defending Honour in Law,' in *Honourable Intentions? Violence and Virtue in Australian and Cape Colonies, c. 1750 to 1850*, ed. Penny Russell and Nigel Worden (London: Routledge, 2016), 17.
17 Nigel Worden, 'Forging a Reputation: Artisan Honour and the Cape Town Blacksmith Strike of 1752,' *Kronos* 28, no. 1 (2002): 44.
18 Arlette Farge, *The Allure of the Archives* (New Haven, CT: Yale University Press, 2013), 91.
19 T.G. Parsons, 'Was John Boston's Pig a Political Martyr? The Reaction to Popular Radicalism in Early New South Wales,' *Journal of the Royal Australian Historical Society* 71, no. 3 (1985): 164–76. See also D.R. Hainsworth, *The Sydney Traders: Simeon Lord and his Contemporaries* (Melbourne: Melbourne University Press, 1981).
20 Boston v Laycock, NSW Court of Civil Jurisdiction; Despatch from Governor Hunter to Duke of Portland, August 1796, *Historical Records of Australia*, Series 1, Vol. 1 (1796): 602–43, accessed 25 May 2018, www.austlii.edu.au/cgi-bin/sinodisp/au/cases/nsw/NSWSupC/1796/5.html. Unless otherwise specified, all details of the case below are drawn from this source.
21 Bruce Kercher, *Debt, Seduction and Other Disasters: The Birth of Civil Law in Convict New South Wales* (Sydney: Federation Press, 1996), 26–27.
22 Lisa Ford, 'The Pig and the Peace: Transposing Order in Early Sydney,' in *Law and Politics in British Colonial Thought: Transpositions of Empire*, eds. Shaunnagh Dorset and Ian Hunter (New York: Palgrave Macmillan, 2010), 169.
23 Ford, 'The Pig and the Peace,' 169.
24 Ford, 'The Pig and the Peace,' 174–76.

25 Ford, 'The Pig and the Peace,' 177.
26 Ford, 'The Pig and the Peace,' 177–80.
27 Kercher, *Debt, Seduction and Other Disasters*, 27.
28 Parsons, 'John Boston's Pig,' 171, 173.
29 Peter N. Moogk, '"Thieving Buggers" and "Stupid Sluts": Insults and Popular Culture in New France,' *William and Mary Quarterly* 36, no. 4 (1979): 526.
30 Ann-Carita Evaldsson, 'Staging Insults and Mobilizing Categorizations in a Multiethnic Peer Group,' *Discourse and Society* 16, no. 6 (2005): 763–86.
31 Jane Austen, *Sense and Sensibility* (London: R. Bentley, 1833), 276–87.
32 *Sydney Gazette,* 29 June 1814.
33 *Sydney Gazette,* 28 September 1816.
34 Keyword search for 'rascal' and/or 'scoundrel' in NSW newspapers, 1803–1850, https://trove.nla.gov.au/newspaper/. Accessed 25 May, 2018.
35 Sandra Blair, 'The Felonry and the Free? Divisions in Colonial Society in the Penal Era,' *Labour History* 45 (1983): 3.
36 See John Hirst, *Convict Society and its Enemies: A History of New South Wales* (Sydney: Allen & Unwin, 1983); Kirsten McKenzie, *Scandal in the Colonies: Sydney and Cape Town, 1820–1850* (Melbourne: Melbourne University Press, 2004).
37 Robert Shoemaker, 'The Decline of Public Insult in London 1660–1800,' *Past and Present* 169, no. 1 (2000): esp. 111–12.
38 Ex parte Salomon (1834), Supreme Court of NSW, accessed 25 May 2018, www.austlii.edu.au/cgi-bin/sinodisp/au/cases/nsw/NSWSupC/1834/97.html.
39 *Sydney Gazette,* 4 July 1833.
40 *Sydney Herald,* 4 July 1833.
41 Honour was nevertheless evoked as the sticking point in multiple cases of interpersonal violence, as Catie Gilchrist has shown. See Catie Gilchrist, 'Defending Honour in Australian Settler Society,' in *Honourable Intentions? Violence and Virtue in Australian and Cape Colonies, c. 1750 to 1850*, ed. Penny Russell and Nigel Worden (London: Routledge, 2016), 140–54.
42 Blair, 'Felonry and Free,' 3–4.
43 Kirsten McKenzie, 'Of Convicts and Capitalists: Honour and Colonial Commerce in 1830s Cape Town and Sydney,' *Australian Historical Studies* 33, no. 118 (2002): 209–13. See also *Sydney Herald,* 24 October 1836.
44 *Sydney Monitor,* 2 April 1831.
45 *Sydney Monitor,* 18 July 1832. On Edward Smith Hall, see Erin Ihde, *A Manifesto for New South Wales: Edward Smith Hall and the* Sydney Monitor, *1826–1840* (Melbourne: Australian Scholarly Publishing, 2004).
46 On Wilson and the scandalous accusations and charges of corruption he faced at this time, see James Bradley and Cassandra Pybus, 'From Slavery to Servitude: the Australian Exile of Elizabeth and Constance,' *Journal of Australian Colonial History* 9 (2007): esp. 39–43.
47 *Sydney Monitor,* 30 January 1839.

Bibliography

Primary sources

Printed Sources:

Austen, Jane. *Sense and Sensibility*. London: R. Bentley, 1833.
Sydney Gazette.
Sydney Herald.
Sydney Monitor.

Legal Proceedings:

Atkins v Harris. NSW Court of Civil Jurisdiction Proceedings. State Records NSW 2/8150 (1799). Accessed 25 May 2018. www.austlii.edu.au/cgibin/sinodisp/au/cases/nsw/NSWSupC/1799/4.html.

Boston v Laycock. NSW Court of Civil Jurisdiction; Despatch from Governor Hunter to Duke of Portland, August 1796. *Historical Records of Australia*, Series 1, Vol. 1 (1796): 602–43. Accessed 25 May 2018. www.austlii.edu.au/cgibin/sinodisp/au/cases/nsw/NSWSupC/1796/5.html.

Ex parte Salomon. Supreme Court of NSW (1834). Accessed 25 May 2018. www.austlii.edu.au/cgi-bin/sinodisp/au/cases/nsw/NSWSupC/1834/97.html.

Secondary sources

Atkinson, Alan. 'Writing about Convicts: Our Escape from the One Big Gaol.' *Tasmanian Historical Studies* 6, no. 2 (1999): 17–28.

Bailey, Peter. 'Breaking the Sound Barrier.' In *Hearing History: A Reader*, edited by Mark M. Smith, 23–35. Athens: University of Georgia Press, 2004.

Blair, Sandra. 'The Felonry and the Free? Divisions in Colonial Society in the Penal Era.' *Labour History* 45 (1983): 1–16.

Bradley, James and Cassandra Pybus. 'From Slavery to Servitude: The Australian Exile of Elizabeth and Constance.' *Journal of Australian Colonial History* 9 (2007): 29–50.

Damousi, Joy. *Colonial Voices: A Cultural History of English in Australia*. New York: Cambridge University Press, 2010.

Damousi, Joy. '"The Filthy American Twang": Elocution, the Advent of American "Talkies," and Australian Cultural Identity.' *American Historical Review* 111, no. 2 (2007): 394–416.

Erlmann, Veit. 'But What of the Ethnographic Ear? Anthropology, Sound, and the Senses.' In *Hearing Cultures: Essays on Sound, Listening and Modernity*, edited by Veit Erlmann, 1–20. Oxford: Berg, 2004.

Evaldsson, Ann-Carita. 'Staging Insults and Mobilizing Categorizations in a Multiethnic Peer Group.' *Discourse and Society* 16, no. 2 (2005): 763–86.

Farge, Arlette. *The Allure of the Archives*. New Haven, CT: Yale University Press, 2013.

Ford, Lisa. 'The Pig and the Peace: Transposing Order in Early Sydney.' In *Law and Politics in British Colonial Thought: Transpositions of Empire*, edited by Shaunnagh Dorset and Ian Hunter, 169–86. New York: Palgrave Macmillan, 2010.

Gilchrist, Catie. '"A Life of Noisy Riot, of Filth, Indecency and Profaneness": The Convict Voice and the Bourgeois Imagination.' *Journal of the Royal Australian Historical Society* 92, no. 1 (2006): 29–45.

Gilchrist, Catie. 'Defending Honour in Australian Settler Society.' In *Honourable Intentions? Violence and Virtue in Australian and Cape Colonies, c. 1750 to 1850*, edited by Penny Russell and Nigel Worden, 140–54. London: Routledge, 2016.

Hainsworth, D.R. *The Sydney Traders: Simeon Lord and his Contemporaries*. Melbourne: Melbourne University Press, 1981.

Hirst, John. *Convict Society and its Enemies: A History of New South Wales*. Sydney: Allen & Unwin, 1983.

Ihde, Erin. *A Manifesto for New South Wales: Edward Smith Hall and the Sydney Monitor, 1826–1840.* Melbourne: Australian Scholarly Publishing, 2004.

Kamensky, Jane. *Governing the Tongue: The Politics of Speech in Early New England.* New York: Oxford University Press, 1997.

Kercher, Bruce. *Debt, Seduction and Other Disasters: The Birth of Civil Law in Convict New South Wales.* Sydney: Federation Press, 1996.

Laugesen, Amanda. *Convict Words: Language in Early Colonial Australia.* Melbourne: Oxford University Press, 2002.

McKenzie, Kirsten. 'Defining and Defending Honour in Law.' In *Honourable Intentions? Violence and Virtue in Australian and Cape Colonies, c. 1750 to 1850*, edited by Penny Russell and Nigel Worden, 17–30. London: Routledge, 2016.

McKenzie, Kirsten. 'Of Convicts and Capitalists: Honour and Colonial Commerce in 1830s Cape Town and Sydney.' *Australian Historical Studies* 33, no. 118 (2002): 199–222.

McKenzie, Kirsten. *Scandal in the Colonies: Sydney and Cape Town, 1820–1850.* Melbourne: Melbourne University Press, 2004.

Moogk, Peter N. '"Thieving Buggers" and "Stupid Sluts": Insults and Popular Culture in New France.' *William and Mary Quarterly* 36, no. 4 (1979): 524–47.

Parsons, T.G. 'Was John Boston's Pig a Political Martyr? The Reaction to Popular Radicalism in Early New South Wales.' *Journal of the Royal Australian Historical Society* 71, no. 3 (1985): 164–76.

Russell, Penny. 'Throwing down the Gauntlet: Voice, Power and Sexual Violence in Penal New South Wales.' In *A Cultural History of Sound, Memory and the Senses*, edited by Joy Damousi and Paula Hamilton, 93–108. London: Routledge, 2017.

Shoemaker, Robert. 'The Decline of Public Insult in London 1660–1800.' *Past and Present* 169, no. 1 (2000): 97–131.

Smith, Mark M., ed. *Hearing History: A Reader.* Athens: University of Georgia Press, 2004.

Smith, Mark M. *Listening to Nineteenth-Century America.* Chapel Hill: University of North Carolina Press, 2001.

Worden, Nigel. 'Forging a Reputation: Artisan Honour and the Cape Town Blacksmith Strike of 1752.' *Kronos* 28, no. 1 (2002): 43–65.

Afterword

Useful echoes

Mark M. Smith

This splendid collection invites conversation. Precisely because it is brimming with persuasive arguments and full of ideas that will pique the attention of historians generally, sensory historians for all periods and places specifically, I think it most helpful if, here, I do not reiterate at any length what Peter Denney, Bruce Buchan, David Ellison and Karen Crawley do in their masterful introduction to the collection. Rather, I would like to ponder some of the ideas inspired by the essays. In other words, I would like to think aloud about some of the collection's useful echoes.

To my mind, there are many such echoes or points of elaboration, but I light on just three, namely: how these incisive essays about the changing meanings and experiences of sound during the long eighteenth century in Britain (from roughly 1700 to the mid-nineteenth century) help us think more deeply about the multi- and intersensory nature of the long Enlightenment; how this collection might help us think about alternative – if complementary – ways to periodise sensory history; and, lastly, how the collection might invite us to think about other ways sound and civility braided. Specifically, I want to think about how, in the period under consideration, war studies could help us expand our understanding of sound, civility and the eighteenth century.

Sound, Space and Civility in the British World maintains that noise increasingly became a signature of modernity, one resonating, literally and figuratively, in burgeoning cities, in increased numbers of voices and in emerging industrialism. As the various essays also amply demonstrate, Enlightenment, Romantic and Victorian commentators came to understand – and, in turn, culturally impose – a definition of sound that stood in contrast to erratic and uncontrollable noise. In other words, they anchored sound to civility. Noise was associated with the hoi polloi and the unrefined tongue befitting the coarse ear. This much has been demonstrated by any number of scholars interested in historical acoustemology generally. What is less well understood – and now, thanks to this collection, is becoming much clearer – is how, despite the obvious increase in noise in this period, it was not always perceived as injurious. As several essays show, contemporaries understood the usefulness of noise in anchoring important abstractions. Lawfulness, industriousness and faith, among many others, could now be understood sonically and situated socially.

Nestled among these trenchant arguments is the observation that, as Denney, Buchan, Ellison and Crawley note in their introduction, much of sensory history has recently been dedicated to examining the multi-sensory creation of space and identity in the long eighteenth century. They rightly point to William Tullett's article on olfaction, but there is other work also worth considering not least because it tends to support the larger suggestion offered in the essays, namely that all of the senses were keenly indexed to notions of civility.[1]

Take, for example, Carolyn Purnell's recent study, *The Sensational Past: How the Enlightenment Changed the Way We Use Our Senses*. The fundamental aim of the book is to show that the sense of sight was not wholly constitutive of truth and knowledge in a period traditionally understood as ocularcentric. Purnell, rather, wants us to understand the interconnection and mutual participation of the senses. The state of technology alone ensured that sight was hardly the master of the senses. This was, after all, a time before widespread street lighting and, as such, activities in, for example, public markets were guided as much by sound and touch as by eyes that struggled in the near dark conditions. Candles were expensive and night workers – such as baker boys (known as 'bats') – learned to rely on their other senses, most notably touch. Patrons frequenting the markets also found sight less than reliable when it came to buying food. Smell and touch were the more trusty sentinels here; the murk of the markets meant that the colour and health of fruits and vegetables simply could not be seen. Hearing also augmented squinting eyes when trying to navigate the markets. Sounds, calls, shouts and noises helped map space and ascertain prices.[2]

Just as sound and noise demarcated social worth in the British eighteenth-century world, Purnell shows how the senses more generally were also intimately tied to inner worth and character. Enlightenment consumers ate food not simply for its nutritional value; taste could also influence personality, emotions and intelligence. Coffee, for example, was considered sober liquor, stimulating creativity without inviting drunkenness. Voltaire supposedly drank 40 cups of it every day. Olfaction played an ambivalent role in French Enlightenment thinking. At once associated with the uncivilised that needed their sense of smell to communicate, it was also considered the sensory organ in closest contact with the brain and therefore connected to intelligence and the imagination. It was, in fact, indispensable to scientific and medical practice especially.

All of this is to say that the authors in *Sound, Space and Civility in the British World* are on to something important. If Purnell is right to highlight the ways in which all of the senses informed Enlightenment thinking, this edited collection of essays helps us understand, precisely, how one sense worked to inform those ways at multiple levels. Read broadly, sound and noise, habits of listening and hearing, braided with smells and habits of smelling, tastes and ways of tasting, sights and ways of seeing, and tactility and methods of touching to texture the experience of daily life. *Sound, Space and Civility in the British World* explains in unprecedented detail how sounds were constitutive of difference and similarity, of incivility and civility, and were not simply injurious but also empowering. These essential points are present but muted in Purnell's otherwise engaging work; this collection animates the point in superb detail. It is one that should be born in mind in future treatments of what might now reasonably be called the sensory Enlightenment.

My second point concerns periodisation. The editors are entirely correct to point to the relative scarcity of work on this period.[3] I take their point seriously and would like to use it to suggest something about periodisation and the senses more generally. In some important respects, one reason why this period – the long Enlightenment, if you will – has received rather less attention than, say, the ancient, early modern and modern periods is probably that sensory history does not lend itself terribly well to conventional historical periodisation. Why? Principally because we are dealing with the longue durée when it comes to the senses; sensory change was often slow, incremental echoes from earlier periods lingering and braiding with new sensory stimuli and modes of perception in often quiet, subtle fashion. Despite the manifest cultural and technological changes apparent in the period 1700–1850, the debates over noise and sound, listening and hearing, detailed by these essays were actually part of a much longer trajectory. As several scholars of the early modern period and even before – including the medieval and ancient world – have shown, discussions about who was noisy and who was sound were driven by elites for hundreds of years with social standing mattering a great deal in designating who was noisy (and noisome) and who understood the intellectual propriety of silence and modulated sound.[4] The period covered in this volume, c.1700–1850, is difficult to characterise sensorially simply because many of the tensions apparent here had existed, albeit in germinated form with different constituencies, much earlier. But if we reframe the history of the senses generally in terms of political economy, the 1700–1850 designation begins to make good sense. This is the era of classical political economy, a period when state intervention into the public sphere to regulate noise and sound was present but, compared to later years, quite modest.

The 1850 cut-off also makes good sense, at least in the Anglo-American world, because it serves as a precursor to the rise of another form of political economy, one centred around greater state intervention in the economy and society generally. In Anglo-America, this is the period of progressivism, when local municipalities especially began to exert greater control over the definition of noisemaking. Here, the state intervened to better regulate and tether the connection between noise and civility.[5]

Although we certainly need more work on twentieth-century soundscapes, especially for the post-1945 period, and most especially for the very recent era with the rise of so-called neo-liberalism and the associated retreat from state regulation (usually dated as beginning in the 1980s), it seems like a safe guess that the rise of manifest state intervention in Western European countries between and after the two World Wars, was likely reflected in the ways that government increasingly attempted to control the modern soundscape, especially with regard to traffic, industry and the noise pollution in the skies.[6] The precise nature of the connection demands much more detailed analysis (not to mention the pressing need for interrogation of the relationship between neo-liberalism and very late twentieth-century political economies), but it might well be the case that rather than index the history of the senses generally – sound included – to extant and quite finely sliced periodisations (Enlightenment, industrialisation, and the like), we could equally talk about the political economy of the senses in broader terms. By treating such a broad period of time (1700–1850), this collection helps us think about what a reperiodised history of the senses might look and sound like. This is a welcome intervention and one that will hopefully inspire others to think about how best to periodise sensory histories.

Lastly, I wish to ponder very briefly, and by way of conclusion, the relationship between sound and what we might think of as manifest incivility in this period by talking about the most obvious example: war. Bruce Buchan helpfully identifies the importance of war in his excellent essay and I would like to build on his keen insight. It is worth remembering that wars littered Europe in the eighteenth century. There were dozens of conflicts and rebellions: some small, some massive in scale; some involving imperial conquest, others contained within borders. Recently, scholars of sound and historical acoustemology have begun to listen to war in some detail in an effort to divine not only the experience of war but also the meaning of the particular conflict itself. More generally, historians of the senses have begun to ask how the larger context defines the meaning of particular wars.[7] What did war sound like – and why – in the long eighteenth century? Did the sounds of muskets, of horses or of armies and navies add to the soundscapes described in this collection? Did the civilian population hear war and, if so, what did they hear and how did they listen to and understand those noises, those sounds, alien and perhaps terrifying? Did these sounds serve as aural counterpoint to the (relatively civil) sounds of peace? Were sounds of war constitutive of imperial acoustemologies? Was the American Revolution heard as a war for liberty, the Seven Years' War as a war for empire? In other words, could and did the sounds of war in the long eighteenth century redefine what was civil and uncivil sound and noise during times of peace, as

Buchan suggests? As yet, we have no fully developed answers to these questions, but in offering us these trenchant essays detailing the nature of civility and sound, this collection surely allows us to begin to think about how to properly frame and understand possible answers.

Notes

1 William Tullett, 'The Macaroni's "Ambrosial Essences": Perfume, Identity and Public Space in Eighteenth-Century England,' *Journal for Eighteen-Century Studies* 38, no. 2 (2015): 163–80. See also Andrew Kettler, 'Odor and Power in the Americas: Olfactory Consciousness from Columbus to Emancipation,' PhD Dissertation, University of South Carolina, 2017; Mark M. Smith, 'Preface: Styling Sensory History,' *Journal for Eighteenth-Century Studies* 35, no. 4 (2012): 469–72.
2 Carolyn Purnell, *The Sensational Past: How the Enlightenment Changed the Way We Use Our Senses* (New York: Norton, 2017).
3 The editors point to the work that has been written. Here I add, in addition to Purnell's book, James H. Johnson, *Listening in Paris: A Cultural History* (Berkeley: University of California Press, 1996).
4 See, for example, Shane Butler and Sarah Nooter, eds., *Sound and the Ancient Senses* (New York: Routledge, 2018); C.M. Woolgar, *The Senses in Late Medieval England* (New Haven, CT: Yale University Press, 2006), 63–83; Bruce R. Smith, *The Acoustic World of Early Modern England: Attending to the O-Factor* (Chicago: University of Chicago Press, 1999); Mark M. Smith, ed., *Hearing History: A Reader* (Athens: University of Georgia Press, 2004), 69–84.
5 Smith, *Hearing History*, 319–30; John M. Picker, *Victorian Soundscapes* (New York: Oxford University Press, 2003).
6 See, especially, James Mansell, *The Age of Noise in Britain: Hearing Modernity* (Urbana: University of Illinois Press, 2017); Karen Bijsterveld, *Mechanical Sound: Technology, Culture and Public Problems of Noise in the Twentieth Century* (Cambridge, MA.: MIT Press, 2008); Daniel Morat, ed., *Sounds of Modern History: Auditory Cultures in 19th and 20th Century Europe* (New York: Berghahn, 2014).
7 See, for example, Mark M. Smith, *The Smell of Battle, The Taste of Siege: A Sensory History of the Civil War* (New York: Oxford University Press, 2014); Mark M. Smith, 'Krieg,' in *Handbuch Sound*, eds. Daniel Morat and Hansjakob Ziemer (Stuttgart: J. B. Metzler, 2018), 391–95; and Paul Cornish and Nicholas J. Saunders, eds., *Modern Conflict and the Senses* (New York and London: Routledge, 2017).

Bibliography

Bijsterveld, Karen. *Mechanical Sound: Technology, Culture and Public Problems of Noise in the Twentieth Century*. Cambridge, MA: MIT Press, 2008.
Butler, Shane and Sarah Nooter, eds. *Sound and the Ancient Senses*. New York: Routledge, 2018.
Cornish, Paul and Nicholas J. Saunders, eds. *Modern Conflict and the Senses*. New York and London: Routledge, 2017.
Johnson, James H. *Listening in Paris: A Cultural History*. Berkeley: University of California Press, 1996.

Kettler, Andrew. 'Odor and Power in the Americas: Olfactory Consciousness from Columbus to Emancipation.' PhD Dissertation, University of South Carolina, 2017.

Mansell, James. *The Age of Noise in Britain: Hearing Modernity*. Urbana: University of Illinois Press, 2017.

Morat, Daniel, ed. *Sounds of Modern History: Auditory Cultures in 19th and 20th Century Europe*. New York: Berghahn, 2014.

Picker, John M. *Victorian Soundscapes*. New York: Oxford University Press, 2003.

Purnell, Carolyn. *The Sensational Past: How the Enlightenment Changed the Way We Use Our Senses*. New York: Norton, 2017.

Smith, Bruce R. *The Acoustic World of Early Modern England: Attending to the O-Factor*. Chicago: University of Chicago Press, 1999.

Smith, Mark M., ed. *Hearing History: A Reader*. Athens: University of Georgia Press, 2004.

Smith, Mark M. 'Krieg.' In *Handbuch Sound*, eds. Daniel Morat and Hansjakob Ziemer. Stuttgart: J. B. Metzler, forthcoming 2018.

Smith, Mark M. 'Preface: Styling Sensory History.' *Journal for Eighteenth-Century Studies* 35, no. 4 (2012): 469–72.

Smith, Mark M. *The Smell of Battle, The Taste of Siege: A Sensory History of the Civil War*. New York: Oxford University Press, 2014.

Tullett, William. 'The Macaroni's "Ambrosial Essences": Perfume, Identity and Public Space in Eighteenth-Century England.' *Journal for Eighteen-Century Studies* 38, no. 2 (2015): 163–80.

Woolgar, Christopher Michael. *The Senses in Late Medieval England*. New Haven, CT: Yale University Press, 2006.

Index